Understanding Cancer Therapies

T0187455

Edited by
Prakash Srinivasan Timiri Shanmugam
Department of Biochemistry and Molecular Biology
Louisiana State University Health Sciences
Center–Shreveport
Shreveport, Louisiana

CRC Press
Taylor & Francis Group
Boca Raton London New York

CRC Press is an imprint of the
Taylor & Francis Group, an **informa** business

CRC Press
Taylor & Francis Group
6000 Broken Sound Parkway NW, Suite 300
Boca Raton, FL 33487-2742

First issued in paperback 2020

© 2018 by Taylor & Francis Group, LLC
CRC Press is an imprint of Taylor & Francis Group, an Informa business

No claim to original U.S. Government works

ISBN 13: 978-0-367-65733-8 (pbk)
ISBN 13: 978-1-138-19815-9 (hbk)

This book contains information obtained from authentic and highly regarded sources. While all reasonable efforts have been made to publish reliable data and information, neither the author[s] nor the publisher can accept any legal responsibility or liability for any errors or omissions that may be made. The publishers wish to make clear that any views or opinions expressed in this book by individual editors, authors or contributors are personal to them and do not necessarily reflect the views/opinions of the publishers. The information or guidance contained in this book is intended for use by medical, scientific or health-care professionals and is provided strictly as a supplement to the medical or other professional's own judgement, their knowledge of the patient's medical history, relevant manufacturer's instructions and the appropriate best practice guidelines. Because of the rapid advances in medical science, any information or advice on dosages, procedures or diagnoses should be independently verified. The reader is strongly urged to consult the relevant national drug formulary and the drug companies' and device or material manufacturers' printed instructions, and their websites, before administering or utilizing any of the drugs, devices or materials mentioned in this book. This book does not indicate whether a particular treatment is appropriate or suitable for a particular individual. Ultimately it is the sole responsibility of the medical professional to make his or her own professional judgements, so as to advise and treat patients appropriately. The authors and publishers have also attempted to trace the copyright holders of all material reproduced in this publication and apologize to copyright holders if permission to publish in this form has not been obtained. If any copyright material has not been acknowledged please write and let us know so we may rectify in any future reprint.

Except as permitted under U.S. Copyright Law, no part of this book may be reprinted, reproduced, transmitted, or utilized in any form by any electronic, mechanical, or other means, now known or hereafter invented, including photocopying, microfilming, and recording, or in any information storage or retrieval system, without written permission from the publishers.

For permission to photocopy or use material electronically from this work, please access www.copyright.com (http://www.copyright.com/) or contact the Copyright Clearance Center, Inc. (CCC), 222 Rosewood Drive, Danvers, MA 01923, 978-750-8400. CCC is a not-for-profit organization that provides licenses and registration for a variety of users. For organizations that have been granted a photocopy license by the CCC, a separate system of payment has been arranged.

Trademark Notice: Product or corporate names may be trademarks or registered trademarks, and are used only for identification and explanation without intent to infringe.

Library of Congress Cataloging-in-Publication Data

Names: Shanmugam, Prakash Srinivasan Timiri, editor.
Title: Understanding cancer therapies / editor, Prakash Srinivasan Timiri Shanmugam.
Description: Boca Raton : Taylor & Francis, 2018.
Identifiers: LCCN 2017041420 | ISBN 9781138198159 (hardback : alk. paper)
Subjects: | MESH: Neoplasms--therapy
Classification: LCC RC270.8 | NLM QZ 266 | DDC 616.99/406--dc23
LC record available at https://lccn.loc.gov/2017041420

Visit the Taylor & Francis Web site at
http://www.taylorandfrancis.com

and the CRC Press Web site at
http://www.crcpress.com

Contents

Acknowledgments

It is a great pleasure to acknowledge our contributors for sharing their knowledge and providing evidence-based information on cancer treatment in our text, *Understanding Cancer Therapies*.

I wish to personally thank the writers for their contributions to our inspiration and knowledge and for providing other help in creating this book.

I am very grateful to our families for their understanding and unrelenting support and encouragement that allowed us to spend extra time on completing the book.

We owe our sincere thanks to the staff at CRC Press/Taylor & Francis Group for their patience and assistance at various stages of the publication of this text.

Prakash Srinivasan Timiri Shanmugam

Editor

Prakash Srinivasan Timiri Shanmugam secured his PhD in the specialization of Pharmacology and Environmental Toxicology-Chemistry (Inter-Disciplinary) from University of Madras, Tamil Nadu, India. He completed his postdoctoral research at Tulane University and LSUHSC-Shreveport, Shreveport, Louisiana. His current research focuses on evaluating pharmacokinetic actions as well as toxic and therapeutic mechanisms of cancer treatment drugs. Additionally, in India, he served as lecturer in the Department of Pharmacology for more than 2 years and had 6 months of experience in clinical research.

Timiri Shanmugam has published 11 research articles in various peer-reviewed international journals, three book chapters, and 18 conference proceedings/abstracts, and has attended various workshops and seminars. He has reviewed several research articles.

Contributors

Chirom Aarti
Research Department of Plant Biology
and Biotechnology
Loyola College, University of
Madras
Chennai, India

Paul Agastian
Research Department of Plant Biology
and Biotechnology
Loyola College
Chennai, India

Pramila Bakthavachalam
Centre for Toxicology and Developmental
Research (CEFT)
Sri Ramachandra University
Chennai, India

Sathya Chandran
Orthodontics and Dentofacial
Orthopaedics
SM Dental Care
Chennai, India

Prince Clarance
Research Department of Plant Biology
and Biotechnology
Loyola College
Chennai, India

Bhargavi Dasari
Department of Oral Medicine and
Radiology
Sibar Institute Of Dental Sciences
Guntur, India

and

Preceptorship in Oral and Maxillofacial
Radiology
Rutgers School of Dental Medicine
Newark, New Jersey

Devaraj Ezhilarasan
Department of Pharmacology
Saveetha Dental College and Hospitals
Saveetha University
Chennai, India

Balu Karthika
Department of Oral Medicine and
Radiology
Priyadarshini Dental College and Hospital
Tiruvallur, India

Ameer Khusro
Research Department of Plant Biology
and Biotechnology
Loyola College, University of Madras
Chennai, India

Sowmiya Renjith
Dental Surgeon
Royal Dent
Chennai, India

Nathiya Shanmugam
Department of Pharmacology
Sri Ramakrishna Dental College and
Hospital
Coimbatore, India

Izaz Shaik
Operative Dentistry and Endodontics
Department of Operative Dentistry and
Endodontics
Sibar Institute Of Dental Sciences
Guntur, India

**Prakash Srinivasan Timiri
Shanmugam**
Department of Biochemistry and
Molecular Biology
Louisiana State University Health
Sciences Center–Shreveport
Shreveport, Louisiana

1 Basics of Radiation and Radiotherapy

Prakash Srinivasan Timiri Shanmugam
and Pramila Bakthavachalam

CONTENTS

INTRODUCTION

More than 11 million people worldwide are diagnosed with cancer every year, and it is estimated that this number will increase to 16 million new cases by 2020. Cancer cells grow abnormally, and they can damage neighboring tissues by direct growth (invasion), or they can spread to distant parts of the body through the bloodstream or lymphatic system (metastases). The unregulated growth is induced by damage to DNA that causes mutations in genes that encode proteins controlling cell division. Cancer is treated by surgery, chemotherapy, radiotherapy, and biologic therapy or their combinations.

Radiation biology is the study of the effects of ionizing radiation on living things. In 1895, the physicist Wilhelm Conrad Röntgen discovered "a new kind of ray" that he called x-rays, with the "x" representing the unknown. The first reported medical use of x-rays was to locate a metal fragment in the spine of a drunken sailor who had been temporarily paralyzed by the injury. By 1897, physicians were using x-rays to treat abnormal growths, such as hairy moles, ulcers, and warts. Radiation therapy developed in the early 1900s, largely due to the milestone work of the Nobel Prize-winning Marie Curie, who discovered radioactive polonium and radium.

CLASSIFICATION OF RADIATION

Radiation may be classified as ionizing or nonionizing. Nonionizing radiation does not have enough energy to produce ions in the irradiated matter. Instead of producing charged ions, the nonionizing radiation has adequate energy only for excitations. Examples of nonionizing radiation are ultraviolet radiation, visible light, and infrared radiation. Ionizing radiation has sufficient energy to break chemical bonds and eject electrons from the atoms. Ionizing radiation can be divided into directly ionizing and indirectly ionizing radiation. Directly ionizing particles are charged particles (electrons, protons, and ions) with sufficient energy to disrupt the atomic structure of the matter and produce ions. The indirectly ionizing particles (uncharged particles like neutrons, photons) do not produce chemical and biological changes by themselves, but they can eject directly ionizing charged particles from atoms in matter (Carlsson 1978).

IONIZING RADIATION

Ionizing radiation carries energies high enough to cause ionizations in exposed matter, leading to chemical alterations. Ionizing radiation includes photon-radiation (γ- and x-ray) and particle radiations (α-, ß-particles, neutrons, heavy ions, etc.).

The energies of the different radiation types are measured in electron volt (keV-MeV). Different types of ionizing radiation can also be characterized by their linear energy transfer, which is measured in keV/μm. For example, α-particles have a higher linear energy transfer than γ-radiation. Thus, α-particles have a denser ionization pattern and therefore deposit more energy per micrometer, causing more clustered damages along its path than γ-radiation. Gamma radiation travels deeper into matter than α-particles and has a more scattered ionization pattern (Goodhead 1994).

The absorbed dose, with the unit Gray (Gy), is defined as the energy deposited in a defined mass of material, where 1 Gy is equal to 1 joule/kg. However, the same dose can cause different amounts of damages depending on the radiation quality. To be able to compare the damages caused by radiation with different linear energy transfer, the relative biological effectiveness can be used. Relative biological effectiveness is calculated by dividing the x-ray or γ-radiation dose with the dose of another type of radiation needed to produce the same biological effect. For radiation protection purposes, it is necessary to estimate the cancer risk caused by exposure to different types of radiation. For this purpose, the equivalent dose, measured in Sievert (Sv), is calculated. The equivalent dose takes into consideration a radiation weighting factor (WR), approximating the relative cancer risk from different radiation qualities per unit dose.

To assess the risks associated with different radiation qualities and organ sensitivities, the effective dose is calculated. The effective dose considers the dose, the type of radiation, and the risk factor of the exposed organ(s). Thus, the effective dose is the sum of the equivalent dose for each organ, multiplied with a weighting factor related to the risk for that organ.

PARTICULATE RADIATIONS

Charged Particles

Electrons
Electrons are negatively charged particles. They may be accelerated to high energy (their speed is then close to the velocity of light) using a betatron or linear accelerator. Electrons are widely used for cancer treatment (Hall and Giaccia 2006).

Protons
Protons are positively charged, almost 2,000 times heavier than electrons. Because of their mass, they must be accelerated in larger devices than electrons, so cyclotrons or synchrotrons are used.

α-Particles
These particles consist of two neutrons and two protons bound together with a helium nucleus. Like protons, they can be accelerated in cyclotrons or synchrotrons. The α-particles are ejected from radioactive nuclei (uranium or radium) in an alpha decay. The α-particles are the main source of natural background radiation.

Heavy Charged Particles
Heavy charged particles are ions of elements such as carbon, oxygen, neon, and others. They are positively charged, and they usually have high kinetic energy (~100 MeV).

Neutrons

Neutrons are electrically neutral particles with mass slightly larger than that of a proton. Neutrons for medical uses are usually produced by bombarding a beryllium target with protons.

CELL DEATH BY IONIZING RADIATION

Radiation has two principal modes of effect for interaction with cellular targets: direct effects and indirect effects. Direct effects take place when, for example, a photon directly ionizes deoxyribonucleic acid (DNA) or other cellular components. Indirect effects are mediated through radiolysis of cellular water, which produces reactive oxygen species. Reactive oxygen species are radicals that cause damages to nucleic acid and other cellular components through oxidation. For γ-radiation, approximately 70% of the damage in cells is caused by the indirect effects of radiation.

The DNA in the cell nucleus is considered to be the most important target for radiobiological effects. DNA has a double helix structure with two polynucleotide strands held together by hydrogen bonds between the bases of the nucleotides. The double-ringed nucleobases adenine (A) and guanine (G) pair with the single-ringed thymine (T) and cytosine (C), respectively. The double helix interacts with small protein complexes called *histones* and is arranged in a compact structure called *chromatin*. The structure of chromatin depends on the stage of the cell cycle; the chromatin is loosely structured during replication, whereas it is tightly packed during cell division.

Ionizing radiation causes DNA damage through two mechanisms of action: by directly causing ionization events within the DNA, or by inducing the formation of free radicals that react with the DNA (i.e., in the form of breaks on the strands, either directly by ionizing the DNA or indirectly by forming free radicals that damage the DNA). However, DNA damage is not always lethal to the cell. Single-strand breaks rarely cause cell death in normal cells since they are easily repaired by the cell using the opposite strand as a template. Double-strand breaks, breaks close to each other on both of the strands, are more difficult to repair.

Following exposure to ionizing radiation, cells can undergo apoptosis, mitotic catastrophe, and/or terminal cell arrest. The extent to which one mode of cell death predominates over another is unclear but may be influenced by cell type, radiation dose, and the cell's microenvironment (e.g., relative oxygenation). Depending on the severity of damage, the tumor suppressor protein P53 can trigger cell cycle arrest or initiate apoptosis via transcriptional activation of pro-apoptotic proteins, including those of the Bcl-2 family. P53-induced protein with a death domain, another P53 pro-apoptotic target, also plays a critical role in DNA damage-induced apoptosis, leading to caspase-2 activation and subsequent mitochondrial cytochrome c release.

In cells irradiated with lethal doses that damage the DNA beyond repair, ionizing radiation (IR) can also induce terminal growth arrest leading to a senescent-like morphology. Terminally arrested cells are metabolically active but incapable of division; they eventually die, days to weeks following IR, via necrosis. It is suggested that the terminal-arrest pathway begins with the transactivation of the CDK2 inhibitor p21, which is involved in the initial induction of senescent-associated G1-arrest. Expression of p21 subsequently declines, while stable expression of the CDK4 inhibitor p16INK4A is induced, thereby maintaining this arrest.

Although IR-induced DNA lesions are lethal if left unrepaired, cell membrane damage can also contribute to apoptosis. Radiation-induced cleavage of plasma membrane-localized sphingomyelin by sphingomyelinases results in the rapid formation of ceramide, a lipid second messenger that is a potent inducer of apoptosis. Subsequent activation of the stress-activated protein kinase (SAPK) signaling cascade via ceramide will then initiate apoptosis.

For the majority of cells, mitotic catastrophe-induced necrosis accounts for most of the cell death following ionizing radiation. Mitotic catastrophe is characterized by abnormal nuclear morphology (e.g., multiple micronuclei or multinucleated giant cells) following premature entry into mitosis by cells manifesting unrepaired DNA breaks and lethal chromosomal aberrations, this often resulting in the generation of nonclonogenic aneuploid and polyploid cell progeny. It is suggested that the abrogation of the G2-M checkpoint is due to overaccumulation of cyclin B and premature activation of the CDK1-cyclin B complex. Radiation-induced mitotic catastrophe is the predominant mode of cell death in P53-deficient tumor cells, which are defective in the G1-S checkpoint and can be selectively arrested by the G2-checkpoint upon DNA damage.

Most of the radiation-induced DNA lesions can be reversed by cellular repair mechanisms. In the repair process, DNA breaks may fail to rejoin correctly and form chromosomal aberrations. Serious chromosome changes hinder correct cell division and lead to mitotic death. High linear energy transfer radiation is believed to cause damage that is more difficult to repair compared to low linear energy transfer radiation due to the dense ionization pattern, which causes multiple strand breaks close in space to a higher degree (Goodhead 1994; Karlsson and Stenerlöw 2004). Consequently, the ability of high linear energy transfer radiation to cause nonrepairable DNA damage is less dependent on alterations, such as mutations affecting the cell's repair capacity, cell cycle phase, and environmental conditions such as oxygen pressure.

The repair mechanisms of cells in fast proliferating tissue such as tumors generally have an increased likelihood of failure. It is therefore advantageous to partition the treatment into multiple fractions. The treatment fractions are typically delivered with daily intervals, which is a time-scale that permits the cells in normal tissue to recover from the effects of the irradiation. Fractionated delivery also increases the probability that each tumor cell at some point during the treatment is exposed to radiation when it is in a radiosensitive state. The fraction dose and the number of fractions are determined based on the estimated number of tumor cells and their radiosensitivity.

CELL CYCLE

The radiosensitivity of cells varies considerably as they pass through the cell cycle. It seems to be a general tendency for cells in the S phase to be the most resistant and for cells in G2 and mitosis (M) to be the most sensitive. The reason for the resistance in S phase is thought to be homologous recombination, increased as a result of the greater availability of an undamaged sister template. A sister chromatid is used as a template to faithfully recreate the damaged section and join the ends together properly. Sensitivity in G2 probably results from the fact that those cells have little time to repair radiation damage before the cell is called upon to divide:

- In general, cells are most radiosensitive in the M and G2 phases, and they are most resistant in the late S phase.
- The cell cycle time of malignant cells is shorter than that of some normal tissue cells, but during regeneration after injury, normal cells can proliferate faster.
- Cell death for nonproliferating (static) cells is defined as the loss of a specific function, while for stem cells it is defined as the loss of reproductive integrity (reproductive death). A surviving cell that maintains its reproductive integrity and proliferates indefinitely is said to be clonogenic (Podgorsak 2003).

DNA DAMAGE

The number of ionizations produced at therapeutic dose levels is very high—approximately 10^5 ionizations per cell per Gray—but the vast majority of these produce no cytotoxic damage. The biological effect of ionizations is influenced by three main factors: free-radical processes, the number of ionizations that are close enough to DNA to damage it, and cellular repair processes.

DNA is the most important cellular component that can be stricken by radiation because it carries the genetic code. The most sensitive part of the DNA molecule is considered to be the pyrimidine bases (T, C). Other important molecules that can be damaged by radiation are the enzymes and the peptide bond in proteins of the cellular membranes. Radiation induces many kinds of lesions to the DNA molecule. The cell succeeds in repairing many of the lesions, while some others are transmitted to posterities, and a small portion leads to cell death. The most common damages of the DNA molecule are presented in Table 1.1.

The DNA double-strand break (DSB) is therefore thought to be the most important type of cellular damage. Just one residual DSB may be sufficient to produce a

TABLE 1.1
Common Damages of the DNA Molecule

Type of Damage	Number per Gray
Double-strand break (DSB)	50
Single-strand break (SSB)	100–500
Base decay	1000–2000
Sugar decay	800–1600
Cross-conjunction DNA-DNA	30
Cross-conjunction DNA-proteins	150

significant chromosome aberration and thus to sterilize the cell. However, since cells have the ability to repair some of those damages, not all the DSB leads to cell death. According to Ward (1986), DSB damage is lethal when it exists together with other DSB damages or with SSB damages.

Other than lymph cells, progametes, and serous cells that after irradiation are destroyed during the S phase, all kinds of mammalian cells undergo "mitotic death". During mitotic death, cells do not die immediately but rather when they attempt mitosis. Even though RNA synthesis is affected by radiation, there is still some amount of RNA that can compose normal protein, and so the cell function can continue for a while.

CHROMOSOMAL DAMAGE

Ionizing radiation can cause alterations in chromosomes by several modes as listed below:

Genetic mutations: These are alterations in the genetic code that can cause cell death or the altered genetic code can be transmitted to the posterities.

Quantitative alterations: Quantitative alterations of the cellular DNA can be made, and multipeptide gigantic cells can be created.

Morphological chromosomal alterations: This morphological weathering is visible during cell mitosis. Chromosome aberrations occur when a cell is irradiated in the early stages of interphase. Interphase is the time when the cell is growing and copying its chromosomes in preparation for cell division. G1 and G2 are stages of the interphase. Radiation causes several segments of the chromosomes to break. These segments tend to adhere to other parts of the defective chromosomes, not to unimpaired ones, resulting in chromosomes with uncommon morphology. Sometimes, these segments reunify to the correct vacancies of the stricken chromosomes, causing the damage to be repaired. It is also possible to have no segment-to-chromosomes reunion at all, and so a chromosome with an eliminating part occurs.

The above impairments usually induce cell death during mitosis, and they are called *unstable* lesions. Other chromosomal impairments, which concern the minority, are called *stable* lesions. Stable lesions are more hazardous because they can be transferred to posterity.

If a cell is going to be irradiated early in the cell cycle, before the DNA synthesis phase occurs, the damage may concern the whole chromosomes, because the chromosomes have not yet started to redouble, causing the radiation to be more effective. However, if the cell is irradiated later in the cell cycle, each chromosome has already started to divide into two chromatids, so the resulting damage occurs in one out of two chromatids for each chromosome.

There are two main groups of abnormalities: chromosome aberration and chromatid aberration. Chromosome aberration results from irradiating cells early in interphase (G1 phase), before genetic material has been duplicated (Hall and Giaccia 2006). Irradiation of cells later in interphase, after the DNA has doubled and the chromosomes consist of two strands of chromatin (G2 phase), leads to chromatid damage. A break that occurs in a single chromatid arm leaves the other arm of the same chromosome undamaged. Irradiation in the S phase can lead either to chromatid aberration or to chromosome aberration.

Many types of chromosome abnormalities are possible, not necessarily associated with cell inactivation.

LETHAL ABERRATIONS

Dicentric or tricentric: This refers to an interchange between two separate chromosomes.

Formation of a ring: A break is caused by irradiation in each arm of a single chromatid early in the cell cycle, and the sticky ends may form a ring and a separate fragment.

Anaphase bridge: Breaks occur in both chromatids of the same chromosome, and sticky ends rejoin incorrectly. In anaphase, when the two sets of chromosomes move to opposite poles of the cell, the region of chromatin between the centromeres is stretched across between the poles, and separation into new daughter cells is inhibited.

NONLETHAL REARRANGEMENTS

Symmetric translocation: Radiation induces breaks in two different chromosomes since the replication has begun, and the broken pieces are exchanged between these two chromosomes.

Small interstitial or terminal deletion: One or two breaks in the same arm of the same chromosome lead to the loss of genetic information in the end of the chromosome arm or between the two breaks.

REACTIVE OXYGEN SPECIES (ROS)

ROS are formed by the indirect effect of irradiation, where 1 Gy of γ-radiation is estimated to result in approximately 1,500,000 ROS per cell. ROS include superoxide anion ($O_2^{\bullet-}$), hydroxyl radical ($\bullet OH$), and hydrogen peroxide (H_2O_2).

ROS are also produced constantly by cellular metabolism. Approximately 2×10^{10} molecules of $O_2^{\bullet-}$ and H_2O_2 per cell per day are formed during cellular respiration (in rat cells). The majority of endogenously produced ROS are derived from the mitochondria and peroxisomes. This production leads to approximately 104 DNA modifications per cell per day. Thus, ROS are highly reactive and can oxidize cellular components such as nucleic acid.

Since ROS are continually produced in the cell, a number of defense mechanisms (belonging to the cellular antioxidant capacity) have evolved to reduce the damage caused by ROS. Oxidative stress is defined as the stage when ROS levels exceed the capacity of the antioxidant defense systems, causing increased levels of ROS. During oxidative stress, the damage to cellular targets such as nucleic acids in DNA or in the nucleotide pool increases and gives rise to different base modifications, for example, 8-oxo-7,8-dihydroguanine (8-oxo-Gua) (in DNA) or

8-oxo-7,8-dihydro-2′-deoxyguanosine-5′triphosphate (8-oxo-dGTP) (in the nucleotide pool). All nucleotides in the 2′-deoxyribonucleoside triphosphate pool can be oxidized by ROS.

MODERN HYPOFRACTIONATED TECHNIQUES

Technological advances in radiation therapy techniques in the 1980s and 1990s brought image-based treatment planning, multileaf collimators, and novel ways to make sure the patient and the tumor were correctly positioned on the treatment table. This enabled us to create and deliver dose distributions with a considerably decreased volume of irradiated tissue outside the tumor. Highly conformal dose distributions and better fractionation sensitivity are now creating new possibilities for hypofractionated treatments. Two techniques that make use of this are stereotactic body radiation therapy and brachytherapy.

STEREOTACTIC BODY RADIATION THERAPY

Stereotactic radiation therapy is an external beam radiation therapy technique that is utilized to irradiate small tumors or lesions. The treated lesions are preferably small, the patient is carefully set up in the dedicated system, and a multileaf collimator with small leaf width is usually used. The resulting dose distributions are exceptionally conformal, but the treatment setup procedure is time consuming. The use of fewer fractions is thus both a possibility and a practical necessity. Stereotactic radiation therapy was initially used to treat intracranial tumors and nonmalignant, malfunctioning vascular bundles.

Stereotactic body radiation therapy was introduced in the 1980s. It uses similar principles as stereotactic radiation therapy but is applied outside the cranium (Blomgren 1995). So far, most experience in stereotactic body radiation therapy has been with liver and lung tumors. The rationale behind hypofractionated liver and lung stereotactic body radiation therapy, besides the poor outcomes for conventional fractionation, was that both organs were considered to tolerate high doses in small volumes (Lax 1994).

BRACHYTHERAPY

Brachytherapy is also called internal radiation therapy. It is based on placing a radioactive material directly inside or next to the tumor volume (Mazeron 2005). The greatest advantage of this type of treatment is the possibility of using a higher total dose of radiation to treat a smaller area. In addition, it is faster than external radiation treatment. There are two types of brachytherapy: temporary and permanent. In the case of temporary brachytherapy, the radioactive material is placed inside or near a tumor for a limited time and then removed. Temporary brachytherapy can be realized as a low-dose rate or high-dose rate treatment. Permanent brachytherapy is also called seed implantation. It involves placing radioactive seeds or pellets (about the size of several millimeters) in or near the tumor and leaving them in place permanently. The time it takes to deliver a treatment varies greatly, from months in permanent implantations with iodine (125 I) or palladium (103 Pd) sources to minutes in high-dose rate brachytherapy with sources of cobalt (60 Co) or iridium (192 Ir). The radioactivity level of the implants gradually decreases and after several months diminishes. The seeds then remain in the body, with no persisting effect on the patient.

OTHER TYPES OF RADIATION

CONVENTIONAL THERAPY

Conventional radiotherapy is based on the delivery of photons or electrons in two-dimensional beams using a linear accelerator to shrink or destroy a tumor. It is usual to deliver the complete radiation dose in many fractions. This procedure spares normal tissues because of the repair of sublethal damage between dose fractions and repopulation of cells (Hall and Giaccia 2006). The damage of tumor cells increases, because of reoxygenation and the reassortment of cells into radiosensitive phases of the cycle between dose fractions.

THREE-DIMENSIONAL CONFORMAL RADIOTHERAPY

Three-dimensional conformal radiation therapy is a modern technique where the multiple x-ray beams are shaped exactly to the contour of the tumor volume. By using computed tomography treatment planning to image and model the tumor and its vicinity in three dimensions, 3D conformal radiation therapy allows normal tissue to be spared much better than does classical irradiation.

INTENSITY-MODULATED RADIATION THERAPY (IMRT)

IMRT is an advanced mode of 3D conformal radiotherapy. It uses specialized software and computer-controlled x-ray accelerators to model the intensity of radiation delivered to the treatment volume. Treatment is planned by using 3D computed tomography (CT) images of the patient, followed by dose calculations to choose the radiation intensity pattern that will best cover the tumor shape. The shaping is achieved by combinations of several intensity-modulated fields coming from different beam directions. Because the dose delivered to normal tissue is significantly lower in comparison to conventional techniques, markedly higher and therefore more effective doses can be delivered to treatment volume with fewer side effects.

THE SIGNIFICANCE OF RADIOTHERAPY IN THE MANAGEMENT OF CANCER

Surgery, which of course has the longer history, is in many tumor types the primary form of treatment, and it leads to good therapeutic results in a range of early non-metastatic tumors.

Radiotherapy has replaced surgery for the long-term control of many tumors of the head and neck, cervix, bladder, prostate, and skin, in which it often achieves

TABLE 1.2
Treatment of Common Side Effects of Radiation

Side Effect	Treatment
Skin erythema or desquamation	Steroid (e.g., 1% hydrocortisone) cream, silver sulfadiazine cream
Mucositis or esophagitis	Saline or sodium bicarbonate oral rinses, codeine or morphine elixir, oral viscous xylocaine, oral mycostatin suspension, oral sucralfate suspension
Nausea and vomiting	Dimenhydrinate or prochlorperazine, ondansetron, oral steroids (e.g., dexamethasone)
Diarrhea	Loperamide, diphenoxylate hydrochloride–atropine sulfate
Cystitis	Phenazopyridine hydrochloride
Proctitis	Hydrocortisone acetate–zinc sulfate, phenylephrine hydrochloride

Source: Samant R, Chuen Chiang Gooi A. *Canadian Family Physician.* 2005;51:1496–1501.

reasonable probability of tumor control with good cosmetic results. In addition to these examples of the curative role of radiation therapy, many patients gain valuable palliation by radiation.

The modern linear accelerator has become a precise tool, capable of depositing a defined dose to a specific volume of tissue, rendering radiotherapy an increasingly important modality for the treatment of most cancers. This has been made possible by rapid advances in technology, including intensity modulation and image guidance in real time. These developments have made it possible to spare normal tissues lying in close proximity to tumors, such as bowel adjacent to cancer of the prostate.

Chemotherapy is the third most important treatment modality at the present time. Following the early use of nitrogen mustard during the 1920s, it has emerged to the point where upwards of 30 drugs are available for the management of cancer, although no more than 10–15 are in common use. Many patients receive chemotherapy at some point in their management, often resulting in symptom relief and disease arrest.

SIDE EFFECTS

Radiotherapy not only kills the cancer cells, but it also affects and/or kills normal cells. Most radiotherapy can be tolerated by many patients with limited side effects. Acute local effects after irradiation therapy are treatable. Chronic side effects are dangerous and should be managed by oncologists. Hair loss is the most common side effect, but it may be temporary (Table 1.2).

CONCLUSION

Radiotherapy is the most important curative and palliative treatment for cancer. Knowledge about the basics of radiotherapy can help patients as they face the difficulties of radiotherapy treatment.

REFERENCES

Blomgren H, Lax I, Naslund I, Svanstrom R. Stereotactic high dose fraction radiation therapy of extracranial tumors using an accelerator: Clinical experience of the first thirty-one patients. *Acta Oncol.* 1995;34(6):861–870.

Carlsson GA. Basic concepts in dosimetry. A critical analysis of the concepts of ionizing radiation and energy imparted. *Radiat Res.* 1978;75(3):462–470.

Goodhead DT. Initial events in the cellular effects of ionizing radiations: clustered damage in DNA. *Int J Radiat Biol.* 1994;65(1):7–17.

Hall EJ, Giaccia AJ. *Radiobiology for the Radiobiologist.* Philadelphia, PA: Lippincott Williams and Wilkins; 2006.

Karlsson KH, Stenerlöw B. Focus formation of DNA repair proteins in normal and repair-deficient cells irradiated with high-LET ions. *Radiat Res.* 2004;161(5):517–527.

Lax I, Blomgren H, Näslund I, Svanström R. Stereotactic radiotherapy of malignancies in the abdomen: Methodological aspects. *Acta Oncol.* 1994;33(6):677–683.

Mazeron JJ. Brachytherapy: A new era. *Radiother Oncol.* 2005;74(3):223–225.

Podgorsak EB. *Review of Radiation Oncology Physics: A Handbook for Teachers and Students.* Vienna, Austria, 2003.

Samant R, Chuen Chiang Gooi A. Radiotherapy basics for family physicians: Potent tool for symptom relief. *Can Fam Physician.* 2005;51:1496–1501.

Ward JF. Mechanisms of DNA repair and their potential modification for radiotherapy. *Int J Radiat Oncol Biol Phys.* 1986;12:1027–1032.

2 Current Trends of Drugs Acting on Oral Squamous Cell Carcinoma (OSCC)

Sowmiya Renjith and Sathya Chandran

CONTENTS

INTRODUCTION

Oral cancer is a highly common malignant tumor of the oral cavity and the 10th most frequent common cancer in the world. The most frequent malignancy is oral squamous cell carcinoma (OSCC), which constitutes more than 90% of the oral malignancies and which has poor prognosis due to therapy-resistant locoregional recurrences and distant metastases (Rajendran and Sivapathasundharam 2012). OSCC is defined as a neoplastic disorder in the oral cavity and is a complex malignancy, where environmental factors, viral infections, and genetic alterations most likely interact and thus give rise to the malignant condition. Development of oral cancer proceeds through epigenetic alteration and discrete molecular genetic changes that are acquired from the loss of genomic integrity after continued exposure to environmental or dietary risk factors (American Cancer Society 2010).

Evidence suggests that oral cancer derives from genetic damage. There are increased risks of oral cancer associated with exposure to genetic mutagens in tobacco, alcohol, and betel quid. Genetic mutations have been detected in oral SCC in chromosomes 3p, 4q, 6p, 8p, 9p, 11q, 13q (retinoblastoma [Rb] tumor suppressor gene), 14q, 17p (p53 tumor suppressor gene), 18q (deleted in colon cancer [DCC] tumor suppressor gene), and 21q. The genetic hypothesis predicts a role for hyperactive oncogenes (growth promoting genes) in oral carcinogenesis. Oncogenes encode many of the signal-transmitting proteins (e.g., EGFr, ras, cytoplasmic kinases, c-myc) via which cells respond to external growth signals. Normal cells, with normal

13

oncogenes, will not commit themselves to another round of DNA replication and cell division without stimulation from such external signals. However, with oncogene mutation, the mutant oncoprotein may send a growth-stimulatory signal to the nucleus, regardless of events taking place in the cell's surroundings. The subsequent autonomous proliferation of mutant oncogene-bearing cells results in tumor formation (Krämer and Löffler 2016).

OSCC is characterized by serial epigenetic and genetic alteration. The accumulations of this alteration leads to uncontrolled cell proliferation of the mutated human oral squamous cells, and the accumulation of damaged genetic material leads to uncontrolled division of mutant oral keratinocytes cells. Multiple gene alterations result in oral carcinogenesis and cause aberrant expression and function of proteins in a number of cellular processes, including apoptosis and angiogenesis.

Natural phytochemicals have received significant interest for chemoprevention and treatment for a wide range of diseases, including OSCC. Phytochemicals, such as polyphenols, are known for their antioxidant capacity and free-radical scavenging properties. In recent years, phytochemicals have received significant attention and been proved to interfere in key cellular pathways. Epigallocatechin-3-gallate (EGCG) is the most abundant and most active phenolic constituent of green tea and has strong antioxidant properties, possessing chemotherapeutic and chemopreventive roles (Chen et al. 2011).

Recently, the involvements of EGCG in apoptotic or autophagy-induced cell death have been increasingly appreciated. Consequently, our main interest was to understand the role of EGCG-mediated cell death, comprising both apoptotic and nonapoptotic cell death, but also the impact on the gene expression pattern for the main genes involved in apoptosis, in SSC-4 cells, a relevant model for oral cancer.

EGCG reduces cell proliferation via apoptosis in a dose-dependent manner, as was confirmed in multiple cell lines. Autophagy is another important mechanism adding to apoptotic processes to reduce cell proliferation (Chen et al. 2011). Important roles in exerting biological active properties are the EGCG oxidation products, such as quinones and semiquinones. The therapeutic actions of EGCG on the human squamous tumor cells involve not only apoptosis but also autophagy. This fact might have significant importance for chemoresistance mechanisms, based on the fact that EGCG is able to target multiple death pathways. It was proved that the autophagy enhances EGCG-induced cell death—this suggests the utility of autophagy inhibitors in enlarging the therapeutic response or in preventing activation of resistance to therapy. EGCG has excellent potential for treatment or as adjuvant therapy for patients with OSCC, by inducing cell death via apoptosis and autophagy (Yu et al. 2016).

METFORMIN

Metformin, which is widely used for patients with type 2 diabetes, may reduce cancer risk. A comprehensive systematic review and meta-analysis performed by Decensi et al. (2010) showed that metformin was associated with a 30% reduction in cancer incidence in individuals with type 2 diabetes compared to patients receiving other diabetic treatments. Promising trends were also observed in overall cancer mortality for patients with pancreatic and hepatocellular cancer and, to a lesser extent, for patients with colon, breast, and prostate cancers. Metformin is an old drug that is now gaining increasing attention as an anticancer agent (Madera et al. 2015). It is thought that metformin exerts anticancer effects through inhibition of insulin and mammalian target of rapamycin (mTOR) pathways. Studies have found that mTOR plays a vital role in the control of cell growth, metabolism, and proliferation, and mediates the phosphoinositide 3-kinase/Akt signaling pathway, which is frequently deregulated in human cancers. AMPK activation results in downregulation of mTOR complex 1 (mTORC1), which is one of the two signaling complexes formed by mTOR and the insulin-like growth factor 1/Akt pathway (Wang et al. 2015). Metformin primarily works by blocking a step in the aerobic production of the cellular energy molecule adenosine triphosphate, which activates a signaling pathway that involves the sensing of cellular energetic stress by the enzyme AMPK AMP-activated protein kinase. Though metformin activates AMPK, it has been shown to work independently of this enzyme. Recent discoveries involving two tumor suppressor proteins, liver kinase B1 (LKB1) and ataxia telangiectasia mutated (ATM), help explain the mechanism by which metformin mediates activation of AMPK (Klein et al. 2016). LKB1 is a well-recognized tumor suppressor that functions as an upstream regulator of AMPK. Its ability to activate AMPK might explain, at least in part, the ability of LKB1 to act as a tumor suppressor. LKB1 may also be an upstream kinase for other members of the AMPK-like subfamily of protein kinases. Similarly to LKB1, ATM is also a tumor suppressor protein. ATM is involved in DNA repair and cell cycle control. In response to metformin, ATM may phosphorylate LKB1, therefore mediating the phosphorylation that activates AMPK. Alternatively, ATM might activate AMPK independently of LKB1, or reduce blood glucose levels through pathways entirely independent of AMPK (Sussman et al. 2011) (Figure 2.1).

The cohort studies by Rêgo et al. (2015) showed that individual patients taking metformin had decreased rates of locoregional recurrence and neck metastasis and improved overall survival and disease-free survival rates compared to individuals not taking metformin. Furthermore, the incidence of head and neck squamous cell carcinoma (HNSCC) was lower in individuals taking metformin versus those not taking metformin. Overall survival (OS) was defined by three cohort studies by Rêgo et al. (2015) but with some different peculiarities. Skinner et al. (2012) reported the improvement in OS and 5 years-OS rate in the patients who took metformin. This cohort study was made in patients with head and neck cancer treated with postoperative radiation therapy who used metformin at the time of radiation (Skinner et al. 2012).

The study by Sandulache et al. (2011) examined the OS in three groups: the first group demonstrated the percentage of OS at 2 years in glottic or supraglottic tumors ($p = 0.01$). The second group examined the OS in the proportion of diabetic patient survivors who were taking metformin or not ($p = 0.05$). Finally, in the third group, the OS was measured through the odds ratio of diabetic patients taking metformin or not ($p = 0.09$ and 0.04, respectively) (Sandulache et al. 2011–2013). Yen et al. (2014) built a large retrospective cohort study, and the incidence of head and neck cancer was 0.64 times lower in the metformin user than in the metformin nonuser ($p < 0.01$).

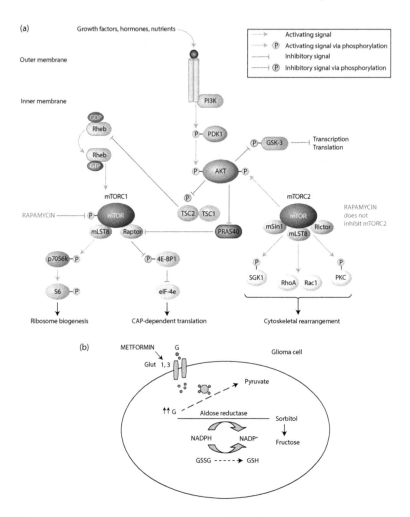

FIGURE 2.1 (a) Pathways of AKT influencing the mTOR protein complexes. Schematic diagram representing the regulatory functions of the mTORC1 and mTORC2 complexes in relation to AKT signaling and cellular outcomes. Induction of AKT activity by extracellular signals results in the activation of mTORC1. mTORC2 activity positively regulates AKT activity. Green arrows represent positive regulation. Green arrows leading to phosphorylation represent activation via phosphorylation. Red arrows represent negative regulation. Red arrows leading to phosphorylation represent repression via phosphorylation. (b) Metformin enhances the activity of glucose transporters (Glut1/Glut3), thus increasing the intracellular concentration of glucose. While glucose has a high affinity for hexokinase, the affinity for aldose reductase is low. As a result, in the presence of high glucose concentrations, hexokinase becomes saturated and the polyol pathway becomes activated. The excess glucose is metabolized by the enzyme aldose reductase to sorbitol, a polyol, and eventually to fructose, in a reaction that uses NADPH (the reduced form of nicotinamide dinucleotide phosphate) as a cofactor. NADPH is also required by the enzyme glutathione reductase in a reaction that regenerates reduced glutathione (GSH). This can compromise the recycling of glutathione disulphide (GSSG) to glutathione. "G" stands for G-type protein. ([a] From Sussman MA, et al. *Physiol Rev.* 2011;91(3):1023–1070.)

There was no significant difference in overall survival between patients with diabetes in the metformin user and metformin nonuser who subsequently developed head and neck cancer ($p = 0.11$). However, the overall survival for patients with diabetes but without head and neck cancer was significantly higher in the metformin user than those in the metformin nonuser ($p < 0.01$) (Yen et al. 2014).

The study by MacKenzie et al. (2012) included an analysis of temsirolimus and metformin in patients with advanced solid tumors in which the patient experienced partial response. The case-control study, which was performed in the United Kingdom, analyzed the relationship between antidiabetic drugs and the risk of head and neck cancer. The authors concluded that antidiabetic drugs were not associated with HNSCC risk, but their data suggested that long-term metformin use had a protective effect against laryngeal cancer (MacKenzie et al. 2012). Vitale-Cross et al. (2012) showed that metformin reduced the growth of HNSCC cells and diminished their mTORC1 activity by both AMPK-dependent and AMPK-independent mechanisms. They also found that metformin specifically inhibited mTORC1 in the basal proliferating epithelial layer of oral premalignant lesions. Remarkably, Vitale-Cross et al. (2012) found that metformin prevented the development of HNSCC. Sandulache et al. (2011) demonstrated that inhibition of respiration by metformin increased glycolytic dependence in wild-type TP53 expressing cells and potentiated the effects of glycolytic inhibition on radiation toxicity. These two works showed the potential clinical benefits of using metformin as a targeted chemopreventive and chemotherapy agent in the control of HNSCC development and treatment (Sandulache et al. 2011).

Most HNSCCs exhibit a persistent activation of the PI3K–mTOR signaling pathway. A study has shown that metformin, an oral antidiabetic drug that is also used to treat lipodystrophy in HIV-infected (HIV+) individuals, diminishes mTOR activity and prevents the progression of chemically induced experimental HNSCC premalignant lesions. They explored the preclinical activity of metformin in HNSCCs harboring PIK3CA mutations and HPV oncogenes, both representing frequent HNSCC alterations, aimed at developing effective targeted preventive strategies. The biochemical and biologic effects of metformin were evaluated in representative HNSCC cells expressing mutated PIK3CA or HPV oncogenes (HPV+). The oral delivery of metformin was optimized to achieve clinical relevant blood levels. Molecular determinants of metformin sensitivity were also investigated, and their expression levels were examined in a large collection of HNSCC cases. They found that metformin inhibits mTOR signaling and tumor growth in HNSCC cells expressing mutated PIK3CA and HPV oncogenes, and that these activities require the expression of organic cation transporter 3 (OCT3/SLC22A3), a metformin uptake transporter. Coexpression of OCT3(octamer-binding transcription factor 3) and the mTOR pathway activation marker pS6 were observed in most HNSCC cases, including those arising in HIV+ patients. Activation of the PI3K–mTOR pathway is a widespread event in HNSCC, including HPV− and HPV+ lesions arising in HIV+ patients, all of which coexpress OCT3. These observations may provide a rationale for the clinical evaluation of metformin to halt HNSCC development from precancerous lesions, including HIV+ individuals at risk of developing HPV-associated cancers (Madera et al. 2015).

In summary, this is evidence of a positive association between decrease of HNSCC and metformin use. Currently available evidence suggests an association between metformin and HNSCC. Metformin appears to improve the overall survival of patients with HNSCC.

CRIZOTINIB

Crizotinib is an anticancer drug acting as an ALK (anaplastic lymphoma kinase) and ROS1 (c-ros oncogene 1) inhibitor (Forde and Rudin 2012; Roberts 2013; Sahu et al. 2013), approved for treatment of some non-small cell lung carcinoma (NSCLC) in the United States and some other countries, and undergoing clinical trials testing its safety and efficacy in anaplastic large cell lymphoma, neuroblastoma, and other advanced solid tumors in both adults and children (NIH ClinicalTrials.gov NCT00932451).

Crizotinib is an inhibitor of receptor tyrosine kinases including *ALK*, Hepatocyte Growth Factor Receptor (HGFR, c-Met), and Recepteur d'Origine Nantais (RON). Translocations can affect the *ALK* gene resulting in the expression of oncogenic fusion proteins. The formation of *ALK* fusion proteins results in the activation and dysregulation of the gene's expression and signaling, which can contribute to increased cell proliferation and survival in tumors expressing these proteins. Crizotinib demonstrates concentration-dependent inhibition of *ALK* and c-Met phosphorylation in cell-based assays using tumor cell lines, and also demonstrates antitumor activity in mice bearing tumor xenografts that express EML4- or NPM-*ALK* fusion proteins or c-Met (Kwak et al. 2010; Cui et al. 2011). Crizotinib is a multitargeted small molecule tyrosine kinase inhibitor, which had been originally developed as an inhibitor of the mesenchymal epithelial transition growth factor (c-MET); it is also a potent inhibitor of *ALK* phosphorylation and signal transduction. This inhibition is associated with G1-S phase cell cycle arrest and induction of apoptosis in positive cells *in vitro* and *in vivo*. Crizotinib also inhibits the related ROS1 receptor tyrosine kinase.

But in recent studies it has been shown that all nine HNSCC protein cell lines showed a varying degree of mesenchymal-epithelial transition factor (MET) protein and RNA expression. Increased MET copy number was not present. MET was expressed after irradiation both *in vitro* and *in vivo*. Crizotinib alone inhibited phosphorylation of MET and inhibited cell growth *in vitro* but did not inhibit phosphorylation of downstream signaling proteins: MAPK, AKT, or c-SRC. When combined with radiation *in vitro*, crizotinib demonstrated radiation enhancement in only one cell line. Crizotinib did not enhance the effect of radiation in either UT-SCC-14 or UT-SCC-15 tumors grown as xenografts (Baschnagel et al. 2015).

CETUXIMAB

Cetuximab is an epidermal growth factor receptor (EGFR) inhibitor used for the treatment of metastatic colorectal cancer, metastatic non-small cell lung cancer, and head and neck cancer (HNC). Cetuximab is a chimeric (mouse/human) monoclonal antibody given by intravenous infusion. Epidermal growth factor receptor (EGFR) is commonly overexpressed or constitutively activated in HNC, including OSCC, and is known to contribute to their uncontrolled proliferation, poor prognosis, and survival (Chung et al. 2006). Hence, blocking the EGFR pathway has been regarded as a promising molecular target for HNC (Boeckx et al. 2014). Unfortunately, only 10%–20% of patients with HNC tumors display a favorable response to cetuximab monotherapy, and even the combination of standard chemotherapies with cetuximab treatment prolongs overall survival by a few months because of resistance to EGFR pathway inhibition (Ratushny et al. 2009; Boeckx et al. 2014). Therefore, new therapeutic approaches, including rational combination strategies, are needed to increase the long-term survival of OSCC patients.

Evidence from recent literature suggested nonthermal atmospheric pressure plasma (NTP) as a promising anticancer therapeutic method by inducing growth arrest and cell death in various types of cancer cells (Chang et al. 2014; Kang et al. 2014). Although the mechanisms underlying the anticancer effects of NTP have not been fully elucidated, the biological effects of NTP are known to depend mainly on reactive oxygen/nitrogen species (ROS/RNS), which are generated when cells and fluid are brought into contact with NTP (Guerrero-Preston et al. 2014; Lee et al. 2014). Previously, we demonstrated that NTP induced anticancer effects via ROS generation. However, we also suggested a novel NTP anticancer mechanism other than ROS signaling. Especially in thyroid papillary cancer cells, NTP ameliorated the invasive characteristics of cancer cells via FAK inhibition, which is associated with both cytoskeleton modulation and inhibition of MMPs/uPA system activities. The purpose of the current study was to explore the effect of combination treatment with NTP on migration and invasion, rather than cell death, of cetuximab-resistant OSCC cell lines. Thus, all experiments were conducted at a concentration (10 µg/mL cetuximab) or intensity (1 kV NTP) that does not result in cell death even in combination. Moreover, in this study, each monotherapy or combination therapy alone had no significant effect on HaCaT normal oral epithelial cells (Chang et al. 2014; Kang et al. 2014; Guerrero-Preston et al. 2014; Lee et al. 2014).

Recent data indicate that NTP in combination with cetuximab had effective antitumor effects by inhibiting migration and invasion of cetuximab-resistant cells via suppression of NF-κB signaling, even though they did not have a significant effect on cancer cell viability. This is the first report to deliver an anticancer effect of NTP other than apoptosis in OSCC, along with the molecular mechanism, even though Guerrero et al. suggested the preliminary possibility of nonapoptotic mechanisms without specific mechanistic explanation in human papilloma virus-negative HNC with a minimal effect on the normal adjacent tissue (Guerrero-Preston et al. 2014). Furthermore, the above was the first report of a strategy comprising the combination of NTP with another anticancer agent in cancer cells to alleviate invasive characteristics (migration/invasion) that are closely associated with local advancement of tumor or distant metastasis. In normal cells, quiescent NF-κB is activated by inflammatory stimuli. In most cancers, including HNC, NF-κB is involved in tumorigenesis, tumor maintenance, or tumor progression and is resistant to cytotoxic chemotherapy (Tanaka et al. 2011). The abnormal constitutive activation of NF-κB contributes to malignant progression and resistance to therapy (Xie et al. 2001). All the cetuximab-resistant cell lines and tumor tissues showed significant NF-κB protein content. Furthermore, cetuximab-sensitive cells demonstrated few NF-κB expressions and downregulation of NF-κB using RNA interference, and recovered cetuximab sensitivity on cetuximab-resistant cells.

Many findings provide new insight into the mechanisms of not only resistance toward cetuximab in OSCC but also treatment with NTP in combination with cetuximab. Several p53-inducing anticancer drugs have been reported to induce p53 as well as NF-κB in various cell types (Hwang et al. 2012). In this condition, despite p53-induced apoptosis, NF-κB activation aggravates invasiveness or promotes resistance to apoptosis. Therefore, NTP, which has the ability to repress NF-κB signaling associated with cetuximab resistance while also activating the p53 pathway, might possess greater anticancer efficacy. Locally advanced oral cavity tumors exhibit a significant threat to survival and function, and treatment includes primary surgical resection with adjuvant chemoradiation therapy, or chemoradiation therapy with surgery reserved for salvage (Cripps et al. 2010).

In the primary chemoradiation therapy protocol, combining NTP application on the primary tumor with chemotherapy can be a promising method of increasing the anticancer effect and decreasing the toxicity by reducing the dose of chemotherapeutics because of its synergistic anticancer effect. An NTP combination with cetuximab is believed to lower delayed treatment failure, which is developed by a few portions of cetuximab-resistant cells in the entire tumor mass (Cripps et al. 2010).

Therefore, NTP combination with chemotherapy, especially with cetuximab, in OSCC is a more feasible and promising therapeutic strategy than in any other cancer type, in particular to control local regional failure with selective antitumor capability in neglecting innocent surrounding tissue, because of the anatomically easy to access of this type of tumor's and synergic effects.

TADALAFIL

Tadalafil is a carboline categorized drug used for treating erectile dysfunction, which in turn seems to beneficially modulate the tumor micro- and macro-environments in patients with HNSCC by lowering myeloid-derived suppressor cells (MDSCs) and regulatory T cells (Treg), and increasing tumor-specific CD8+ T cells in a dose-dependent fashion. Phosphodiesterase-5 (PDE5) inhibition can modulate these cell populations, and the PDE5 inhibitor tadalafil can revert tumor-induced immunosuppression and promote tumor immunity in patients with HNSCC.

The immune system of a patient with HNSCC is suppressed by the gathering of myeloid-derived suppressor cells (MDSC) and regulatory T cells (Treg) whose involvement in many malignancies has been associated with a poor prognosis. Preclinical models have shown that the immunosuppressive action of MDSCs and Treg can be overcome by the use of phosphodiesterase-5 (PDE5) inhibitors (Freiser et al. 2013). Specifically, a short course of daily tadalafil treatment is sufficient to significantly (1) reduce MDSCs and Treg systemically at the tumor site, (2) increase the percentage of tumor-specific CD8þ T cells in circulation, and (3) promote the activation of CD8þ T cells at the tumor site (Weed et al. 2015).

The immune system plays a vital role in the progression of HNSCC as initially suggested by the numerous immunologic defects and the expansion of immunosuppressive units (i.e., MDSCs and Treg) both at the tumor site and in the blood (Freiser et al. 2013). Therapeutic manipulation of the immune system and its response, by corollary, may play a highly significant role in the treatment of HNSCC. In humans, the preclinical findings show a beneficial immune-modulatory effect of PDE5 retardation in the tumor bearer (Serafini et al. 2006; Karakhanova et al. 2013): tadalafil more significantly reduced MDSC and Treg numbers in the blood of patients with HNSCC and increased the concentration of tumor-specific CD8þ T cells. The reason for such a depletion of both MDSCs and Treg is still unknown. On the basis of our preclinical data, we can speculate that PDE5 inhibition can downregulate IL4Ra on MDSCs (Serafini et al. 2006), reducing the survival signals that the receptor mediates (Roth et al. 2012). Subsequently, it is possible that the PDE5 blockade stops the positive loop by which MDSCs promote their own recruitment and differentiation (Molon et al. 2012). The tadalafil-mediated reduction of circulating Treg does not seem to be related to a direct action of this drug, because incubation of Treg with physiologically relevant concentrations of tadalafil does not increase their apoptosis in a small number of preliminary experiments performed.

Hence, further study is needed to completely exclude this possibility. Clinical trials have shown in a preclinical model that MDSCs can expand Treg *in vivo* and that PDE5 inhibition can block this process (Serafini et al. 2008). Considering the rapid turnover of circulating Treg in the tumor-bearing host (Vukmanovic-Stejic et al. 2006; Liston and Gray 2014), it is possible that MDSCs' inhibition significantly decreases Treg proliferation without being altered in their elevated apoptotic rate, de facto reducing their frequency in the blood. Alternatively, the reduced Treg frequency in the blood and in the tumor after tadalafil treatment may be explained by altered Treg homing signals due to changes in the tumor macro-environment. Indeed, advantageous changes in the tumor macro-environment are also suggested by the normalization of the CD4:CD8 T-cell ratio and from the data of a similar but independent trial in which patients with HNSCC have been treated for 15 days with tadalafil before treatment.

REFERENCES

American Cancer Society. *Cancer Facts and Figures 2010*. Atlanta, GA: American Cancer Society. Retrieved from http://www.cancer.org/research/cancerfactsfigures/cancerfactsfigures/cancer-facts-and-figures-2010

Baschnagel AM, et al. Crizotinib fails to enhance the effect of radiation in head and neck squamous cell carcinoma xenografts. *Anticancer Res*. 2015;35(11):5973–5982.

Boeckx C, et al. Overcoming cetuximab resistance in HNSCC: The role of AURKB and DUSP proteins. *Cancer Lett*. 2014;354:365–377.

Chang JW, et al. Non-thermal atmospheric pressure plasma inhibits thyroid papillary cancer cell invasion via cytoskeletal modulation, altered MMP-2/-9/uPA activity. *PloS One*. 2014;9:e92198.

Chen PN, et al. Epigallocatechin-3 gallate inhibits invasion, epithelial–mesenchymal transition, and tumor growth in oral cancer cells. *J Agric Food Chem*. 2011;59(8):3836–3844.

Chung CH, et al. Increased epidermal growth factor receptor gene copy number is associated with poor prognosis in head and neck squamous cell carcinomas. *J Clin Oncol.* 2006;24:4170–4176.

Cripps C, Winquist E, Devries MC, Stys-Norman D, Gilbert R. Epidermal growth factor receptor targeted therapy in stages III and IV head and neck cancer. *Curr Oncol.* 2010;17:37–48.

Cui JJ, et al. Structure based drug design of crizotinib (PF-02341066), a potent and selective dual inhibitor of mesenchymal-epithelial transition factor (c-MET) kinase and anaplastic lymphoma kinase (ALK). *J Med Chem.* 2011;54(18):6342–6363.

Decensi A, et al. Metformin and cancer risk in diabetic patients: A systematic review and meta-analysis. *Cancer Prev Res.* 2010;11:1451–1461.

Forde PM, Rudin CM. Crizotinib in the treatment of non-small-cell lung cancer. *Expert Opin Pharmacother.* 2012;13(8):1195–1201.

Freiser ME, Serafini P, Weed DT. The immune system and head and neck squamous cell carcinoma: From carcinogenesis to new therapeutic opportunities. *Immunol Res.* 2013;57:52–69.

Guerrero-Preston R, et al. Cold atmospheric plasma treatment selectively targets head and neck squamous cell carcinoma cells. *Int J Mol Med.* 2014;34:941–946.

Hwang SG, et al. Anti-cancer activity of a novel small molecule compound that simultaneously activates p53 and inhibits NF-κB signaling. *PloS One.* 2012;7:e44259.

Kang SU, et al. Nonthermal plasma induces head and neck cancer cell death: The potential involvement of mitogen-activated protein kinase-dependent mitochondrial reactive oxygen species. *Cell Death Dis.* 2014;5:e1056.

Karakhanova S, et al. Gender-specific immunological effects of the phosphodiesterase 5 inhibitor sildenafil in healthy mice. *Mol Immunol.* 2013;56:649–659.

Klein JD, et al. Metformin, an AMPK activator, stimulates the phosphorylation of aquaporin 2 and urea transporter A1 in inner medullary collecting ducts. *Am J Physiol Renal Physiol.* 2016;310(10):F1008–F1012.

Krämer A, Löffler H. Biologic features of CUP. In: Krämer A, Löffler H, eds. *Cancer of Unknown Primary.* 2016, Springer International Publishing, Switzerland, 27–44.

Kwak EL, et al. Anaplastic lymphoma kinase inhibition in non-small-cell lung cancer. *N Engl J Med.* 2010;363(18):1693–1703.

Lee SY, et al. Nonthermal plasma induces apoptosis in ATC cells: Involvement of JNK and p38 MAPK-dependent ROS. *Yonsei Med J.* 2014;55:1640–1647.

Liston A, Gray DH. Homeostatic control of regulatory T cell diversity. *Nat Rev Immunol.* 2014;14:154–165.

MacKenzie MJ, Ernst S, Johnson C, Winquist E. A phase I study of temsirolimus and metformin in advanced solid tumours. *Invest New Drugs.* 2012;30:647–652.

Madera D, et al. Prevention of tumor growth driven by PIK3CA and HPV oncogenes by targeting mTOR signaling with metformin in oral squamous carcinomas expressing OCT3. *Cancer Prev Res.* 2015;8(3):197–207.

Molon B, Viola A, Bronte V. Smoothing T cell roads to the tumor: Chemokine post-translational regulation. *Oncoimmunology.* 2012;1:390–392.

NIH ClinicalTrials.gov. Clinical trial number NCT00932451. An Investigational Drug, PF-02341066, Is Being Studied In Patients with Advanced Non-Small Cell Lung Cancer with a Specific Gene Profile Involving the Anaplastic Lymphoma Kinase (ALK) Gene.

Rajendran A, Sivapathasundharam B. *Shafer's Textbook of Oral Pathology.* Elsevier, India, 2012.

Ratushny V, Astsaturov I, Burtness BA, Golemis EA, Silverman JS. Targeting EGFR resistance networks in head and neck cancer. *Cell Signal.* 2009;21:1255–1268.

Rêgo DF, et al. Effects of metformin on head and neck cancer: A systematic review. *Oral Oncol.* 2015;51(5):416–422.

Roberts PJ. Clinical use of crizotinib for the treatment of non-small cell lung cancer. *Biologics.* 2013;7:91–101.

Roth F, De La Fuente AC, Vella JL, Zoso A, Inverardi L, Serafini P. Aptamer-mediated blockade of IL4Rα triggers apoptosis of MDSCs and limits tumor progression. *Cancer Res.* 2012;72:1373–1383.

Sahu A, Prabhash K, Noronha V, Joshi A, Desai S. Crizotinib: A comprehensive review. *South Asian J Cancer.* 2013;2(2):91–97.

Sandulache VC, et al. Association between metformin use and improved survival in patients with laryngeal squamous cell carcinoma. *Head Neck.* 2013;36:1039–1043.

Sandulache VC, et al. Individualizing antimetabolic treatment strategies for head and neck squamous cell carcinoma based on TP53 mutational status. *Cancer.* 2012;118:711–721.

Sandulache VC, et al. Glucose, not glutamine, is the dominant energy source required for proliferation and survival of head and neck squamous carcinoma cells. *Cancer.* 2011;117:2926–2938.

Serafini P, et al. Phosphodiesterase-5 inhibition augments endogenous antitumor immunity by reducing myeloid-derived suppressor cell function. *J Exp Med.* 2006;203:2691–2702.

Serafini P, Mgebroff S, Noonan K, Borrello I. Myeloid-derived suppressor cells promote cross-tolerance in B-cell lymphoma by expanding regulatory T cells. *Cancer Res.* 2008;68:5439–5449.

Skinner HD, et al. TP53 disruptive mutations lead to head and neck cancer treatment failure through inhibition of radiation-induced senescence. *Clin Cancer Res.* 2012;18:290–300.

Sussman MA, et al. Myocardial AKT: The omnipresent nexus. *Physiol Rev.* 2011;91(3):1023–1070.

Tanaka K, et al. Oncogenic EGFR signaling activates an mTORC2-NF-κB pathway that promotes chemotherapy resistance. *Cancer Discov.* 2011;1:524–538.

Vitale-Cross L, et al. Metformin prevents the development of oral squamous cell carcinomas from carcinogen-induced premalignant lesions. *Cancer Prev Res (Phila).* 2012;5:562–573.

Vukmanovic-Stejic M, et al. Human CD4þ CD25hi Foxp3þ regulatory T cells are derived by rapid turnover of memory populations in vivo. *J Clin Invest.* 2006;116:2423–2433.

Wang L, et al. The inhibitory effect of metformin on oral squamous cell carcinoma. *Zhonghua Kou Qiang Yi Xue Za Zhi.* 2015;50(6):360–365.

Weed DT, et al. Tadalafil reduces myeloid-derived suppressor cells and regulatory T cells and promotes tumor immunity in patients with head and neck squamous cell carcinoma. *Clin Cancer Res.* 2015;21(1):39–48.

Xie S, et al. Reactive oxygen species-induced phosphorylation of p53 on serine 20 is mediated in part by polo-like kinase-3. *J Biol Chem.* 2001;276:36194–36199.

Yen YC, et al. Effect of metformin on the incidence of head and neck cancer in diabetics. *Head Neck* 2014;37(9):1268–1273.

Yu CC, et al. Suppression of miR-204 enables oral squamous cell carcinomas to promote cancer stem-ness, EMT traits, and lymph node metastasis. *Oncotarget.* 2016;7(15):20180–20192.

3 Phytopharmaceuticals in Cancer Treatment

Prince Clarance and Paul Agastian

CONTENTS

INTRODUCTION

Cancer is a group of diseases that can affect various organs of the body and is characterized by the uncontrolled growth of abnormal cells and invasion into normal tissue. Cancer cells can also spread to other parts of the body and produce new tumors (Kharwar et al. 2011). Cancer remains one of the most predominant illnesses causing death, where each year more than 10 million people are identified worldwide (Saeidnia and Abdollahi 2014), and 7.4 million cancer-related deaths occur every year (Jemal et al. 2011). The aim of most cancer chemotherapeutic drugs is to inhibit the mechanism implied in the cellular division and thus destroy the malignant tumor cells. Cancer therapy is a very difficult task due to nonspecific toxicity and the severe drug resistance of most anticancer drugs (Nygren and Larsson 2003). The disease treatment becomes extremely difficult with the limited number of chemotherapies and their detrimental side effects and the high costs of the drugs. Many existing therapies do not effectively treat certain types of cancers, and multi-drug-resistant tumors aggravate the challenge ahead. The discovery of new natural chemotherapeutic agents is of great relevance.

The extracts of natural products are of great value in the control of malignancies in view of their low cytotoxic activities and drug resistance (Mbaveng et al. 2011). Plants, fungi, endophytic fungi, bacteria, endophytic bacteria, actinomycetes, and marine organisms are rich sources of novel organic compounds with interesting biological activities and a high level of biodiversity. They represent a relatively unexplored ecological source. Their secondary metabolites are particularly active since most of them have interactions with their hosts (Krohn et al. 2007).

NATURAL ANTICANCER AGENTS

The role of natural products as a source for remedies has been realized since ancient times (Farnsworth et al. 1985; Cragg et al. 1997). Despite major scientific and technological breakthroughs in combinatorial chemistry, drugs derived from natural products still make a major contribution to drug discovery in general and cancer therapy in particular. Plants have a long history of use in the treatment of cancer. In his review, Hartwell lists more than 3,000 plant species that have reportedly been used in cancer treatment (Hartwell 1982).

Plants, microorganisms, and more recently marine organisms of various types have traditionally represented a main source of natural anticancer agents (Butler 2005). It is because of the tremendous chemical diversity that is present in these millions of species. For most living organisms, this chemical diversity is gifted by the impact of evolution in the selection and conservation of self-defense mechanisms that characterize the strategies employed to repel or to destroy predators.

Nature is a great source for potential chemotherapeutic agents as well as for lead compounds that have provided the basis and inspiration for the partial or total synthesis of effective new drugs. A large number of drugs in clinical use as anticancer drugs are of natural-product origin, and it has been estimated that approximately 80% of new chemical entities with small-molecule structures introduced during the period from 1950 to 2010 in this field were natural products or were natural-product inspired (Newman and Cragg 2012).

Although combinatorial chemistry, diversity-oriented synthesis, and high-throughput screening (HTS) of large compound libraries are important technologies in the making of new drugs, the role of natural sources in providing new cytotoxics continues to be relevant (Cragg et al. 2009; Cozzi et al. 2004). A new notion that the use of natural-product templates combined with chemical modifications leads to more selective analogs has a better chance of success than the combinatorial approach is gaining acceptance (Avendano and Menendez 2015). The natural-product lead structure will be considerably used as a template for the construction of novel compounds with enhanced biological properties. At least in the anticancer field, "nature has already carried out the combinatorial chemistry," and all we have to do is refine the structures (Mann 2002, p. 147). All these insights have increased the interest in natural products as drug candidates in cancer therapy.

PLANT-DERIVED ANTICANCER AGENTS

Cancer happens as a result of many changes in a variety of genes and signaling processes. Targeting various signaling elements will be a better approach of treatment. Regarding this point, medicinal plants, in both single and multiple applications, can conserve an amplified potential (Saeidnia and Abdollahi 2014). Natural products, particularly as combinatorial approach, have revealed promising clinical models, of which pomegranate, green tea, soy, and tomato paste are some examples (Kumar et al. 2010).

Several plant-derived compounds are successfully employed in cancer treatment (Tables 3.1 and 3.2). One of the best examples is from a 1950 discovery and development of vinca alkaloids, namely, vinblastine and vincristine (Noble 1990). The introduction of the vinca alkaloid vincristine was responsible for an increase in the cure rates for Hodgkin's disease and some forms of leukemia (DeVita et al. 1970). Vinorelbine, the major semisynthetic tubulin-binding vinca alkaloid, was first approved in Europe in 1989 and in the United States in 1994 for the treatment of non-small cell lung carcinoma (NSCLC) and of advanced breast cancer (Facchini and De Luca 2008). Vincristine inhibits microtubule assembly, inducing tubulin self-association into coiled spiral aggregates (Noble 1990), leaving the dividing cancer cell with condensed chromosomes and blocked in mitosis.

Etoposide is a derivative of podophyllotoxin derived from *Podophyllum peltatum* L and *P emodi* L (Newman et al. 2003). Recently podophyllotoxin was isolated from species belonging to the sections *Linum* (Weiss et al. 1975), *Dasylinum*, and *Linopsis*, which represent other alternative sources of podophyllotoxin (Konuklugil 1996). Etoposide is a topoisomerase II inhibitor, stabilizing the enzyme–DNA cleavable complex leading to DNA breaks (Liu 1989).

Paclitaxel (Taxol) represents the taxane family of drugs. They have shown antitumor activity against breast, ovarian, and other tumor types in clinical trials. Paclitaxel stabilizes microtubules, leading to mitotic arrest. Paclitaxel was originally isolated from the inner bark of the Northwest Pacific yew tree, *Taxus brevifolia* (Wani et al. 1971). There have been numerous attempts to derivatize paclitaxel to enhance its bioavailability and to reduce its toxic side effects. One of the most important of these derivatives is docetaxel, a drug with better solubility characteristics than the

TABLE 3.1
Plant-Derived Anticancer Agents

Plant	Compound	Mechanism of Action	Cancer Use	Reference
Catharanthus roseus	Vinblastine, Vincristine, Vindesine-(semisynthetic derivative)	Targets tubulin and microtubules and depolymerization and condensation	Leukemia, lymphoma Breast, lung, germ-cell and renal cancer	Jordan and Wilson (2004)
Podophyllotoxin peltatum, P. emodi	Etoposide phosphate	Inhibitor of the enzyme topoisomerase II	Ewing's sarcoma, lung cancers, testicular cancers, lymphomas, nonlymphocytic leukemias, and glioblastoma	Newman et al. (2003) Liu (1989)
Taxus brevifolia	Paclitaxel (Taxol) Docetaxel (derivative)	Stimulates microtubule polymerization	Ovarian cancer, breast and lung cancers	Wani et al. (1971), Schiff (1979), Altmann and Gertsch (2007)
Camptotheca acuminate	Camptothecin Topotecan, Irinotecan (derivatives)	Inhibiting the intranuclear enzyme topoisomerase-I	Ovarian and small-cell lung cancers	Hsiang et al. (1985)
Cephalotaxus harringtonia var. Drupacea	Homoharringtomine	Inhibition of protein synthesis	Various leukemias	Kantarjian et al. (1996), Zhou et al. (1995)
Combretum caffrum	Combretastatins	Tubulin binding agents, structurally related to colchicine	Active against colon, lung, and leukemia cancers	Ohsumi et al. (1998), Pettit et al. (1987)
Dysoxylum binectariferum	Rohitukine Flavopiridol (derivative)	First cyclin-dependent kinase inhibitor; interfering with the phosphorylation of cyclin-dependent kinases and arrest cell-cycle progression at growth phase G1 or G2	Leukemia, lymphomas, and solid tumors	Kellard (2000), Christian et al. (1997)

(Continued)

TABLE 3.1 (Continued)
Plant-Derived Anticancer Agents

Plant	Compound	Mechanism of Action	Cancer Use	Reference
Aglaila sylvestre	Sylvestrol	Impairs the ribosome recruitment step of translation initiation by affecting the composition of the eukaryotic initiation factor (eIF) 4F complex	Lung and breast cancer cell lines	Cencic et al. (2009), Cragg and Newman (2005)
Raphanus sativus L.	Olomucine Roscovitine (derivative)	Induce the accumulation of tumor suppressor p53 to arrest cells in G1, G2/M phases, suppression of mRNA synthesis	Bone cancer, lung inflammation, salivary gland dysfunction	Cragg and Newman (2005), Jungman and Paulsen (2001), Cicenas et al. (2015)
Mauritiana rugosa M. oenoplia	Betulinic acid	Triggers the mitochondrial pathway of apoptosis in cancer cells	Cytotoxicity against human melanoma cell lines	Cichewitz and Kouzi (2004), Fulda (2008)
Erythroxylum pervillei Baill	Pervilleine A	Receptor-mediated activity—adjuvant in cancer therapy	Selectively cytotoxic against a multi-drug-resistant oral epidermoid cancer cell line (KB-VI) in the presence of vinblastine	Mi et al. (2001), Mi et al. (2003), Chibale et al. (2012)
Aronia melanocarpa	Chlorogenic acid	Mediating the increase of tumor suppressor genes, reducing oxidative stress, and thus damaging the DNA important for the proliferation of cancer cells	Colorectal and colon cancer	Malik et al. (2003), Han et al. (2005), Shukla and Mehta (2015)
Camellia sinensis	Epigallocatechin-3-gallate	Decreased production of MMP-2, MMP-9, and uPA; induction of cell cycle arrest	Prostate, colon, and gastric cancers; skin cancer; and ovarian carcinoma	Taylor and Wilt (1999), Katiyar et al. (2000), Nishikawa et al. (2006)
Gymnema sylvestre	Gymnemagenol	Interaction with drug target proteins of cancer cells protein kinase I and human protein kinase CK-2	High inhibition of proliferation of HeLa cells	Khanna and Kannabiran (2013), Khanna and Kannabiran (2009)

TABLE 3.2

Phytochemicals Used as Cancer Chemopreventive and Treatment Agents

Phytochemical	Plant Source	Cancer Use	Mechanism of Action	Reference
Apigenin	*Moringa peregrina*	Breast and colon cancer	Induces apoptosis, affects leptin/leptin receptor pathway, and induces cell apoptosis in lung adenocarcinoma cell line	El-Alfy et al. (2011), Chung et al. (2007), Bruno et al. (2011)
Curcumin	*Curcuma longa*	Colon, breast cancers, lung metastases, and brain tumors	Regulation of multiple cell signaling pathways including cell proliferation pathway (cyclin D1, c-myc), cell survival pathway (Bcl-2, Bcl-x, cFLIP, XIAP, c-IAP1), caspase activation pathway (caspase-8, -3, -9), tumor suppressor pathway (p53, p21), death receptor pathway (DR4, DR5), mitochondrial pathways, and protein kinase pathway (JNK, Akt, and AMPK)	Bachmeir et al. (2010), Senft et al. (2010), Ravindran et al. (2009)
Crocetin	*Saffron crocus Crocus sativus*	Hepatocellular carcinoma, pancreas, skin, colorectal cancers	Inhibits nucleic acid synthesis	Gutheil et al. (2012), Amin et al. (2011)
Cyanidins	From berries such as grapes, blackberry, cranberry, raspberry, or apples and plums, red cabbage, and red onion	Colon, breast cancers; chemopreventive potential	Inhibits iNOS and COX-2 gene expression in colon cancer cells	Kim et al. (2008)

(Continued)

TABLE 3.2 (Continued)
Phytochemicals Used as Cancer Chemopreventive and Treatment Agents

Phytochemical	Plant Source	Cancer Use	Mechanism of Action	Reference
Fisetin	*Acacia greggii, Acacia berlandieri* Strawberries, apples	Human colon cancer cells and other tumors	Modulates protein kinase and lipid kinase pathways, dual inhibition of PI3K/Akt and mTOR signaling in human non-small cell lung cancer cells by fisetin	Gabor and Eperjessy (1966), Geraets et al. (2009), Khan et al. (2012)
Genistein	*Flemingia vestita*, soybeans	Breast cancer	Tyrosine kinase inhibition by inhibiting DNA topoisomerase II	Markovits et al. (1989), Lopez-Lazaro et al. (2007)
Gingerol	Ginger	Colon, breast, ovarian, and pancreas cancers	Decreases iNOS and TNF-α expression via suppression of IκBα phosphorylation and NF-κB nuclear translocation	Jeong et al. (2009), Lee et al. (2008), Oyagbemi et al. (2010)
Lycopene	Carrots, watermelons, tomato, and red papayas	Prostate cancer, endometrial and colon cancers	Activates cancer preventive enzymes such as phase II detoxification enzymes	Giovannucci et al. (1995)

parent compound (Cragg and Newman 2005). Camptothecin (CPT) is a quinoline alkaloid, a potent inhibitor of the eukaryotic topoisomerase I (Hsiang et al. 1996). It was first isolated from the Chinese ornamental tree *Camptotheca acuminata* Decne (Nyssaceae), was advanced to clinical trials by the National Cancer Institute (NCI) in the 1970s, but was dropped because of severe bladder toxicity (Potmeisel and Pinedo 1995). Some derivatives of CPT such as topotecan and Camptosar (irinotecan) have been approved for use against ovarian, small lung, and refractory ovarian cancers (Demain and Vaishnav 2011). Homoharringtonine isolated from the Chinese tree *Cephalotaxus harringtonia* var. *Drupacea* (Cephalotaxaceae), is another plant-derived agent in clinical use (Itokawa et al. 2005; Powell et al. 1970). It has shown effectiveness against various leukemias (Kantarjian et al. 1996). The principal mechanism of action of homoharringtonine is the inhibition of protein synthesis, blocking cell-cycle progression (Zhou et al. 1995). Combretastatins were isolated from the bark of the South African tree *Combretum caffrum* (Combretaceae). Combretastatin is active against colon, lung, and leukemia cancers, and it is expected that this molecule is the most cytotoxic phytomolecule isolated so far (Ohsumi et al. 1998; Pettit et al. 1987).

ANTICANCER AGENTS FROM MICROORGANISMS

The most attractive sources of medically valued secondary metabolites are produced by microorganisms. Bacteria are well known as significant sources of new pharmaceutical compounds and drugs (Mahajan and Balachandran 2015). Bacterial toxins can be used for tumor destruction, and cancer vaccines can be based on immunotoxins of bacterial origin (Jain 2001). Filamentous fungi are a great source of naturally derived compounds. Numerous bioactives such as mycotoxins and antifungal and anticancer agents have been reported in the literature within the last 100 years (Frisvad et al. 2004). Endophytic bacteria and fungi are earning wide attention for the past decade. Many endophytic bacteria are known for their diverse range of secondary metabolic products including anticancer compounds (Ryan et al. 2008). There has been a major contribution of endophytic fungi to the discovery of anticancer agents. Of the total compounds isolated from endophytic fungi, 57% were novel or were analogs of known compounds (Kharwar et al. 2011). Actinomycetes have been one of the most investigated groups during the past 50 years (Karikas 2010). Marine actinomycetes have attracted great attention since they have developed unique metabolic and physiological capabilities that could make them potential sources of productive compounds with antitumor and other interesting pharmacological properties that are not found in terrestrial microorganisms (Magarvey et al. 2002; Blunt et al. 2006).

ANTICANCER AGENTS FROM BACTERIA

Bacterial toxins, otherwise called *immunotoxins*, are of great demand in cancer therapy. Bacterial toxins are used as oncolytic agents because of their direct effect on tumors and their effective immunomodulating capacity. Bacterial toxins have a ligand or a fragment of an antibody that is connected to a protein toxin. After the ligand subunit binds to the surface of the target cell, the molecule internalizes, and the toxin kills the cell. *Pseudomonas* exotoxin and diphtheria toxin are examples of bacterial toxins used for targeting cancer cells. Immunotoxins have been produced to target hematological malignancies and solid tumors via a wide variety of growth factor receptors and antigens (Pastan and Kreitman 1998) (Tables 3.3).

Other uses of bacteria in cancer therapy are the unaltered pathogenic bacteria as the oncolytic agents, bacteria as sensitizing agents for conventional chemotherapy, bacteria as delivery agents for anticancer drugs, genetically modified bacteria for selective tumor destruction, manipulation of bacterial genes for the coding enzymes for gene-directed enzyme prodrug therapy, and bacterial drugs for gene therapy (Jain 2001).

TABLE 3.3
Bacterial Toxins

Toxin	Description	Example	Reference
Immunotoxins	They contain a ligand (e.g., growth factor, monoclonal antibody or a fragment of an antibody), which is connected to a protein toxin. Once the ligand binds the surface of the target cell, the molecule internalizes the toxin and kills the cell.	*Pseudomonas* exotoxin Diphtheria toxin	Pastan and Kreitman (1998)
Escherichia coli toxins	SLT 1 has been studied for anticancer effect. It binds to specific tumor cells with receptors and kills them by inhibiting protein synthesis. The receptor is CD 77.	Shiga toxin family contains two types of toxins: Stx 1 (SLT1) and Stx 2 (SLT2); in humans, verotoxin receptors by dendritic cells and targeted against ovarian, testicular, breast, and brain cancers	Arab et al. (1999)
Fusion toxins	TGF-α fused with PE 40. PE 40 was further fused to the IL-2 receptor. The fusion protein is therapeutic to IL-2 receptor–related diseases like adult T-cell leukemia.	TGF-α, *Pseudomonas* exotoxin derivatives	Goldberg et al. (1995), Zhang et al. (1995)
Transferrin CRM 107	Used for the treatment of glioblastoma.	Conjugate of human transferin (TS) and a genetic mutant of diphtheria toxin CRM 107	Laske, et al. (1997)
IL-4 fusion toxin	Fusion toxin is administered directly into the tumor. It binds with high affinity to IL-4 receptors, which exist in malignant tumors and do not exist on normal cells.	IL-4 fused with *Pseudomonas* exotoxin	Puri (1999)

THE IDEAL BACTERIAL ANTICANCER AGENT

- It should maintain its toxicity, invasiveness, and tissue lysis. But this should be selective against the tumor and spare the normal tissues.
- It should not have adverse systemic effects.
- It should respond to available antibiotics and not develop resistance.
- It should have a targeted delivery system to maximize the effect on the tumor.

ANTICANCER AGENTS FROM ENDOPHYTIC BACTERIA

The bacteria that live inside the plant tissues without causing any harm to plants are called *endophytic bacteria*. Endophytic bacteria in association with rhizospheric bacteria extend several benefits for the host plants, such as stimulation of plant growth, nitrogen fixation, and resistance to plant pathogens (Mahajan et al. 2014). The bacteria that reside in living plant tissues are not adequately subjected to studies for exploiting their potential in production of natural compounds such as anticancer agents. In 2013 endophytic bacteria were isolated from *Miquelia dntata* Bedd., which is capable of producing camptothecin, a quinoline alkaloid, a potent inhibitor of the eukaryotic topoisomerase I (Shweta et al. 2013).

Endophytic bacteria have been detected in plants that have anticancer properties such as *Catharanthus roseus, Oscimum sanctum, Alovera, Withania somnifera,* and *Murraya konengii* (Joshi and Kulkarni 2015).

ANTICANCER AGENTS FROM FUNGI

Natural products against cancer from plants and bacteria have a greater role in drug discovery. But from the fungal world, investigations of fungal metabolites and their derivatives have not so far led to a clinical lead molecule, though they do indicate promising anticancer activity. Many of these compounds are awaiting human clinical trials in the near future.

The anticancer metabolites of fungi can be grouped into three categories based on the fungal source from which they are derived. They are (1) phytopathogenic (Table 3.4), (2) toxigenic, and (3) nontoxigenic fungi (Evidente et al. 2014).

PHYTOPATHOGENIC FUNGI

CYTOCHALASINS

Cytochalasins are derived from *Pyrenophora* and *Phoma* species of fungi. They are part of a larger group of metabolites called *cytochalasans*, and they incorporate diverse polyketide–amino acid hybrid structures with a wide range of distinctive biological functions (Scherlach et al. 2010). Cytochalasins have the property of binding to the actin filaments and block polymerization and elongation of actin. Cytochalasins can alter cellular morphology, inhibit cell division, and cause apoptosis.

TABLE 3.4
Mechanisms and Properties of Important Metabolites from Phytopathogenic Fungi

Metabolite	Description	Reference
Cotylenin A	It is a plant growth regulator. Along with rapamycin, it effectively inhibits the proliferation of several breast cancer cell lines. The growth arrest of the MCF-7 cells at the G1 phase is induced by the treatment with cotylenin A and rapamycin.	Kasukabe et al. (2008)
Fusicoccin A	Fusicoccin A targets 14-3-3 proteins in cancer cells and promotes isoform-specific expression of 14-3-3 proteins in human gliomas. It is found that these 14-3-3 proteins play a critical role in serine/threonine kinase-dependent signaling pathways through protein–protein interactions with multiple phosphorylated ligands.	Takahashi et al. (2012)
Ophiobolin O	Ophiobolin O induces growth inhibition of human breast cancer MCF-7 cells through G0/G1 cell cycle arrest and reduces the viability of these cells in a time- and dose-dependent manner through activation of apoptotic processes and modifications in JNK (c-Jun NH2-terminal kinase), p38 MAPK (mitogen-activated protein kinase), and ERK (extracellular signal-regulated kinase) activity as well as the reduction of Bcl-2 phosphorylation (Ser70).	Yang et al. (2012)

FUSICOCCANES

Fusicoccanes is a family of metabolites from *Fusicoccum amygdali* and *Drechslera gigantean*. Cotylenin A is a member of the Fusicoccane family that inhibits the growth of various tumor types both *in vitro* and *in vivo* without apparent adverse effects in human xenograft models (Honma et al. 2003). In addition to this, the Fusicocccane family includes fusicoccin A and ophiobolins. Fusicoccin A is an α-D-glucopyranoside of the diterpenoid 5-8-5 ring skeleton. It is the main phytotoxin produced by *Fusicoccum amygdali*, the causative fungal agent of peach and almond canker (Balli et al. 1964).

The subgroup ophiobolins are produced mainly by the phytopathogenic fungi belonging to the genus *Bipolaris*. Ophiobolins reduce seed germination; reduce growth of roots and coleoptiles of wheat seedings; and at the cellular level, they affect membrane permeability, stimulate leakage of ß-cyanine, electrolytes, and glucose from roots; decrease photosynthetic CO_2 fixation; cause respiratory changes; and enhance stomatal opening (Au et al. 2000).

METABOLITES FROM TOXIGENIC FUNGI

BISORBICILLINOIDS

Two metabolites, namely bislongiquinolide and 2′,3′-dihydrotrichodimerol (bisor-bibutenolide) were found with IC50 concentrations <100 mM in the six cancer cell lines analyzed (Balde et al. 2010). These two natural products were isolated from the fungus *Trichoderma citrinoviride*.

Also, 2,3-dihydrotrichodimerol was shown to activate the peroxisome proliferator-activated receptor-Y (PPAR-Y), which exerts major roles in cancer cell biology. This metabolite has also been reported to suppress the production of tumor necrosis factor-α (TNF-α) and nitric oxide in LPS (lipopolysaccharide)-stimulated RAW264.7 cells (Lee et al. 2005).

Sesquiterpene and Eurochevalierine from *Neosartorya pseudofischeri*

Sesquiterpene and eurochevalierine are considered as precursors in the biosynthesis of the potent benzoxazine natural product CJ-12662, a topoisomerase inhibitor. Eurochevalierine exhibits *in vitro* growth-inhibitory activity in various cancer cell lines. It appears that eurochevalierine represents a novel chemical scaffold for the development of anticancer agents effective against cancers unresponsive to traditional therapy with proapoptotic agents (Eamvijarn et al. 2012), because the ester bond linking the indole-containing portion of the molecule with the terpene residue is hydrolytically labile. Future studies will undoubtedly involve synthetic work aimed at resolving the minimum structural requirements in this scaffold (Evidente et al. 2014).

METABOLITES FROM NONTOXIGENIC FUNGI

Tryprostatins A&B from *Aspergillus fumigatus*

These are indole alkaloidal fungal products. Tryprostatin A was first demonstrated to be an inhibitor of the mitogen-activated protein (MAP)-kinase-dependent microtubule assembly and, through the disruption of the microtubule spindle, to specifically inhibit cell cycle progression at the mitotic phase (Usui et al. 1998).

Halenaquinones from *Xestospongia* cf. *Carbonariai* (A Sponge)

Halenaquinone-inhibited epidermal growth factor (EGF) receptor activity (IC50 = 19 μM) (Chang and Geahlen 1992). Inappropriate expression of RAD51 activity may cause tumorigenesis. RAD51 is an essential enzyme for the homologous recombinational repair (HRR) of DNA double-strand breaks. In the HRR pathway, RAD51 catalyzes the homologous pairing between single-stranded DNA (ssDNA) and double-stranded DNA (dsDNA), which is the central step of the HRR pathway. Halenaquinone is a novel type of RAD51 inhibitor that specifically inhibits the RAD51-dsDNA binding (Takaku et al. 2011).

Anticancer Agents from Endophytic Fungi

The anticancer properties of several secondary metabolites from endophytes have been a major attraction in the past. One hundred anticancer compounds belonging to 19 different chemical classes with activity against 45 different cell lines have been

TABLE 3.5
Anticancer Agents from Endophytic Fungi

Endophytic Species	Medicinal Plant	Compound	Activity	Reference
Pestalotiopsis microspora	*Taxus wallichiana*	Taxol	Anticancer	Maheshwari (2006)
Pestalotiopsis microspora	*Taxodium distichum*	Taxol	Anticancer	Li et al. (1996)
Fusarium solani LCPANCF01	*Tylophora indica*	Taxol	Anticancer	Merlin et al. (2012)
Entrophospora infrequens	*Nathapodytes foetida*	Camptothecin	Anticancer	Sanjana et al. (2012)
Fusarium oxysporum	*Catharanthus roseus*	Vinca alkaloids	Leukemia	Kharwar et al. (2008)
				Giridharan et al. (2012)
Phomopsis glabae	*Pongamia pinnata*	PM181110	Anticancer	Verekar et al. (2014)
Chaetomium globosum IFB-E019	*Imperata cylindrical*	Chaetoglobosin U, C, F, E	Cytotoxic	Ding et al. (2006)
Fusarium solani	*Coscinium fenestratum*	Berberine	Anticancer	Diana and Agastian (2013)

isolated from over 50 different fungal species belonging to six different endophytic fungal groups. Of the total compounds isolated from endophytic fungi, 57% were novel or were analogs of known compounds (Ravindra et al. 2011). Table 3.5 presents some examples of anticancer agents from endophytic fungi.

Taxol

Taxol (also known as Paclitaxel) and some of its derivatives are the first major group of anticancer agents isolated that are produced by endophytes. Taxol is a highly functionalized diterpenoid found in each of the world's yew (*Taxus*) species (Suffness 1995). Taxol has a unique mode of action compared to other anticancer agents (Gangadevi and Muthumary 2008). The mode of action of Taxol is to preclude tubulin molecules from depolymerizing during the process of cell division (Schiff and Horowitz 1980). Taxol was first isolated from *Taxus brevifolia* (Wani et al. 1971).

The most common source of Taxol is the bark of trees belonging to the Taxus family including yew trees. Some of the most commonly found endophytes of the world's yews are *Pestalotiopsis* spp. (Strobel et al. 1996), of which *P. microspora* is one of the most commonly isolated species (Strobel 2002). Later it was found that the *P. microspora* isolates obtained from Bald Cypress, other than *Taxus* spp., were producing Taxol (Li et al. 1996). Further, Taxol production has also been noted in *Periconia* sp. (Li et al. 1998) *Seimatoantlerium nepalense* endophytic fungal species (Bashyal et al. 1999) and from the endophytic fungus *Bartalinia robillardoides* (strain AMB-9) (Gangadevi and Muthumary 2008). Merlin et al. (2012) isolated Taxol from the endophytic fungi *Fusarium solani* LCPANCF01 strain isolated from the plant *Tylophora indica* (Burm. f) (Merlin et al. 2012). What is the reason behind the wide distribution of endophytic fungi that make Taxol? Taxol is a fungicide, and the organisms with the most sensitivity to it are plant pathogens such as *Pythium* spp. and *Phytophthora* spp. (Young et al. 1992). These organisms are strong competitors with endophytic fungi for niches within plants. In fact, their sensitivity to paclitaxel is based on their interaction with tubulin in a manner identical to that in rapidly dividing human cancer cells (Schiff and Horowitz 1980). As a result, endophytes may produce Taxol for the protection of the respective host plant from the pathogenic fungi.

CAMPTOTHECIN

Camptothecin is another important anticancer compound, a potent antineoplastic agent that was first isolated from the wood of *Camptotheca acuminate* Decaisne (Nyssaceae) in China (Wall et al. 1966). Later this alkaloid was reported from several plant species such as *Ophiorrhiza* spp. *Ervatamia heyneana, Merrilliodendron megacarpum*, and with a highest yield by *Nothapodytes nimmoniana* (Govindachari and Viswanathan 1972).

Camptothecin

Camptothecin is also isolated from endophytic fungi *Entrophospora infrequens* from the plant *Nathapodytes foetida* (Sanjana et al. 2012). Camptothecin is a pentacyclic quinoline alkaloid that inhibits topoisomerase I, an enzyme involved in DNA replication. This compound acts as an antineoplastic agent and has a cytotoxic effect by inhibiting the dissociation of the DNA–topoisomerase I complex during replication (Pommier 2006; Ling-Hua et al. 2003).

VINCRISTINE

Vincristine, also known as leurocristine, and vinblastine are two alkaloids isolated from *Catharanthus roseus* or *Vinca rosea* belonging to the family Apocyanaceae.

They are used as major drugs in the treatment of leukemia and lymphoma, respectively (Barnett et al. 1978).

Vincristine

Vinca alkaloids

Vinblastine

It is estimated that about 500 kg of leaves are required to produce just 1 gram of purified vincristine (Balandrin and Klocke 1988). As an alternative source, Kharwar et al. (2008) isolated a total of 183 endophytic fungi from *C. Roseus*, of which *Alternaria* sp. and *Fusarium oxysporum* from phloem of the plant material resulted in production of vinca alkaloids (Kharwar et al. 2008).

SCLEROTIORIN

Sclerotiorin is an orange-yellow colored pigment belonging to the azaphilone class of pigments and was originally isolated from *Penicillium sclerotiorum* (Curtin and Reilly 1940). Today it is isolated from many other filamentous fungi. Recently Giridharan et al. (2012) reported that the endophytic fungi *Cephalotheca faveolata* from the healthy leaves of *Eugenia jambolana* Lam. are able to produce sclerotiorin (Giridharan et al. 2012).

Sclerotiorin

Sclerotiorin possesses activities like endothelin receptor binding (Pairet et al. 1995), inhibition of Grb2-Sch interaction blocking the oncogenic Ras signal, and inhibition of lipase (Nam et al. 2000) The study on sclerotiorin showed that it significantly induced cytotoxicity and increased LDH release in HCT–116 cells treated with sclerotiorin (Giridharan et al. 2012).

Verekar et al. (2014) recently discovered a new depsipeptide anticancer compound from *Phomopsis glabae*, an endophytic fungus from the leaves of *Pongamia pinnata* (L), and the active compound isolated is PM181110. The compound exhibited *in vitro* anticancer activity against 40 human cancer cell lines with a mean IC 50 value of 0.089 μM and *ex vivo* efficacy toward 24 human tumor xenografts (Verekar et al. 2014).

A new cytotoxic cytochalasan-based alkaloid named chaetoglobosin U, along with four known analogs, chaetoglobosins C, F, and E, penochalasin A, has been isolated from endophytic fungus *Chaetomium globosum* IFB-E019 from the stem of *Imperata cylindrica* (Ding et al. 2006).

BERBERINE

Diana and Agastian (2013) isolated an endophytic fungus *Fusarium solani* from *Coscinium fenestratum,* an endangered medicinal plant, and screened for berberine production (Diana and Agastian 2013) an active compound for cardioprotective, antidiabetic, anticancer (Sun et al. 2009), antimicrobial (Schmeller et al. 1997), antidiarrheal, anti-arrhythmic, and antitumor activities (Yamamoto et al. 1993).

Berberine

ANTICANCER AGENTS FROM MARINE ORGANISMS

Marine organisms are rich sources of natural products (Pomponi 1999). There are about 3,000 new substances that have been isolated from marine sources that have potential for novel chemical classes (Schweitzer et al. 1991). Many compounds derived from marine organisms have generated interest in structure elucidation, synthesis, and cytotoxicity. According to Rocha et al. (2001), Table 3.6 presents some marine-derived anticancer agents updated with recent informations.

DIDEMNIN B

The first anticancer product from a marine source is a cyclic depsipeptide called didemnin B isolated from the tunicate *Trididemnum solidum* (Rinehart 2000). Didemnin B inhibits protein synthesis and DNA synthesis much more than RNA synthesis and is in general more potent than didemnin A. It inhibits G1 cell cycle progression at nanomolar concentrations by undefined mechanisms (Crews et al. 1994). The preliminary results showed partial activity against non-Hodgkin's lymphoma (Chun et al. 1986).

ET-743

ET-743 is a tetrahydroisoquinoline alkaloid isolated from the marine tunicate *Ecteinascidia turbinata*. ET-743 induces a broad inhibition of activated transcription (Minuzzo et al. 2000). ET-743 acts by selective alkylation of guanine residues in the DNA minor groove. It so differs from other alkylating agents and also interacts with nuclear proteins (Minuzzo et al. 2000).

TABLE 3.6
Some Marine Organism-Derived Anticancer Agents

Compound	Cancer Use	Mechanism of Action	Status
Cytarabine	Leukemia, lymphoma	Inhibition of DNA synthesis	Phase III/IV
Bryostatin 1	Experimental	Activation of PKC (protein kinase C)	Phase I/II
Dolastatin 10 and 15	Breast and liver cancers	Inhibition of microtubules and pro-apoptotic effect	Phase II
Ecteinascidin 743	Experimental	Alkylation of DNA	Phase II
Aplidine	Experimental- human leukemia cells MOLT-4	Inhibition of cell cycle progression	Phase I/II
Halichondrin B	Metastatic breast cancer	Interaction with tubulin	Phase I/II
Discodermolide	Experimental	Stabilization of tubulin	Preclinical
Cryptophycin	Non-small cell lung cancer	Microtubule inhibitor	Phase I/II

CONCLUSION

Nature has been the source of medicine. In the recent past, many natural novel compounds have been isolated and developed for clinical use, particularly from plant sources. It is certain that nature will continue to be a major source of lead molecules for drugs. The drug potentials of endophytes and marine environment are underexplored. It is becoming increasingly evident that the sphere of microorganisms offers vast untapped potential.

Natural products have been an important source of chemotherapeutics for the past 40 years. Natural products isolated from medicinal plants have played a major role in the treatment of cancer. There are more than 270,000 higher plants on the planet, but only a small portion of their uses have been explored. Of the explored, most anticancer drugs have been discovered through random screening of organism collections, but our improved understanding of many of the molecular details of carcinogenesis and evolution makes it possible to develop more efficient strategies. Knowledge in molecular modeling, combinatorial and computational chemistry, and good biology would help in developing the lead molecule efficiently.

ACKNOWLEDGMENT

The authors are thankful to Loyola College-Times of India, Science Research Project (project code 4LCTOI114PBB002) for the funding in research on novel biomolecules from endophytic organisms from endangered medicinal plants.

REFERENCES

Altmann KH, Gertsch J. Anticancer drugs from nature—Natural products as a unique source of new microtubule-stabilising agents. *Nat Prod Rep.* 2007;88:3888–3890.

Amin A, Hamza AA, Bajbouj K, Ashraf SS, Daoud S. Saffron: A potential target for a novel anti-cancer drug against hepatocellular carcinoma. *Hepatology* 2011;54:857–867.

Arab B, Rutka J, Lingwood C. Verotoxin induces apoptosis and the complete, rapid, long-term culmination of human astrocytoma xenografts in nude mice. *Oncol Res.* 1999;11:33–39.

Au TK, Chick WSH, Leung PC. The biology of ophiobolins. *Life Sci.* 2000;67:733–742.

Avendano C, Menendez CJ. *Medicinal Chemistry of Anticancer Drugs.* 2nd ed., Amsterdam: Elsevier Science; 2015.

Bachmeir BE, Mirisola V, Romeo F, Generoso L. Reference profile correlation reveals estrogen-like transcriptional activity of Curcumin. *Cell Physiol Biochem.* 2010;26(3):471–482.

Balandrin MJ, Klocke JA. *Medicinal, aromatic and industrial materials from plants.* In: Bajaj YPS, ed. *Biotechnology in Agriculture and Forestry.* Vols. 4. Berlin, Germany: Springer; pp. 60–103, 1988.

Balde ES, et al. Investigations of fungal secondary metabolites with potential anticancer activity. *J Nat Prod.* 2010;73:969–971.

Balli A, Chain EB, De Leo P, Erlanger BF, Mauri M, Tonolo MA. Fusicoccin: A new wilting toxin produced by Fusicoccum amygdali del. *Nature.* 1964;203:297.

Barnett CJ, et al. Structure-activity relationships of dimeric Catharanthus alkaloids 1. Deacetyl vinblastine amide (vindesine) sulfate. *J Med Chem.* 1978;21(1; January):88–96.

Bashyal B, Li JY, Strobel GA, Hess WM. Seimatoantlerium nepalense, an endophytic Taxol producing coelomycete from Himalayan yew (*Taxus wallichiana*). *Mycotaxon.* 1999;72:33–42.

Blunt JW, Copp BR, Munro MH, Northcote PT, Prinsep MR. Marine natural products. *Nat Prod Rep.* 2006;23:26–78.

Bruno A, et al. Apigenin affects leptin/leptin receptor pathway and induces cell apoptosis in lung adenocarinoma cell line. *Eur J Cancer.* 2011;47:2042–2051.

Butler MS. Natural products to drugs: Natural product derived compounds in clinical trials. *Nat Prod Rep.* 2005;22:162–195.

Cencic R, et al. Antitumor activity and mechanism of action of the Cyclopenta[b]benzo-furan, silvestrol. *PLoS ONE.* 2009;4(4):e5223. http://doi.org/10.137/journal.pone000 5223.

Chang CJ, Geahlen RL. Protein-tyrosine kinase inhibition: Mechanism-based discovery of antitumor agents. *J Nat Prod.* 1992;55:1529–1560.

Chibale K, Davies-Coleman M, Masimirembwa C. *Drug Discovery in Africa: Impacts of Genomics, Natural Products, Traditional Medicines, Insights into Medicinal Chemistry and Technology Platforms in Pursuit of New Drugs.* New York, NY: Springer; 2012.

Christian MC, Pluda JM, Ho TC, Arbuck SG, Murgo AJ, Sausville EA. Promising new agents under development by division of cancer treatment, diagnosis and centers of the National Cancer Institute. *Semin Oncol.* 1997;13:2643–2655.

Chun HG, et al. Suffness. Didemnin B: The first marine compound entering clinical trials as an antineoplastic agent. *Invest New Drugs.* 1986;4:279–284.

Chung CS, Jiyang Y, Cheng D, Birt DF. Impact of adenomatous polyposis coli (APC) tumor suppressor gene in human colon cancer cell lines on cell cycle arrest by apigenin. *Mol Carcin.* 2007;46(9):773–782.

Cicenas J, et al. Roscovitine in cancer and other diseases. *Ann Transl Med.* 2015;3(10):135. Doi:10.3978/j.issn.2305-5839.2015.03.61

Cichewitz RH, Kouzi SA. Chemisry, biological activity and chemotherapeutic potential of betulinic acid for the prevention and treatment of cancer and HIV infection. *Med Res Rev.* 2004;24:90–114.

Cozzi P, Mongelli N, Suarato A. Recent anticancer cytotoxic agents. *Curr Med Chem Anticancer Agents.* 2004;4(2):93–121.

Cragg GM, Newman DJ. Plants as source of anticancer agents. *J Ethnopharmacol.* 2005;100:72–79.

Cragg GM, Newman DJ, Snader KM. Natural products in drug discovery and development. *J Nat Prod.* 1997;60:52–60.

Cragg GM, Grothaus PG, Newman DJ. Impact of natural products on developing new anticancer agents. *Chem Rev.* 2009;109(7):3012–3043.

Crews CM, Collins JL, Lane WS, Snapper ML, Schreiber SL. GTP-dependent binding of the antiproliferative agent didemnin to elongation factor 1 alpha. *J Biol Chem.* 1994;269(22):15411–15414.

Curtin TM, Reilly J. Sclerotiorin, a chorinated metabolic product of *Penicillium sclerotiorum* van Beyama. *Biochem J.* 1940;34:1418–1421.

Demain AL, Vaishnav P. Natural products for cancer chemotherapy. *Microbial Biotechnol.* 2011;4:687–699.

DeVita VT Jr, Serpick AA, Carbone PO. Combination chemotherapy in the treatment of advanced Hodgkin's disease. *Ann Intern Med.* 1970;73:881–895.

Diana VS, Agastian P. Berberine production by endophytic fungus *Fusarium solani* from *Coscinium fenestratum*. *IJBPR.* 2013;4(12):1239–1245.

Ding G, et al. Chaetoglobosin U, a cytochalasan alkaloid from endophytic *Chaetomium globosum* IFB -E019. *J Nat Prod.* 2006;69(2):302–304.

Eamvijarn A, et al. Secondary metabolites from a culture of the fungus *Neosartorya pseudofischeri* and their in vitro cytostatic activity in human cancer cells. *Planta Med.* 2012;78:1767–1776.

El-Alfy TS, Ezzat SM, Hegazy AK, Amer AM, Kamel GM. Isolation of biologically active constituents from *Moringa peregrina* (Forssk.) Fiori. (family: Moringaceae) growing in Egypt. *Pharmacogn Mag.* 2011;7(26):109–115.

Evidente A, et al. Fungal metabolites with anticancer activity. *Nat Prod Rep.* 2014;31:617–627.

Facchini PJ, De Luca V. Opium poppy and Madagascar periwinkle: Model non-model systems to investigate alkaloid biosynthesis in plants. *Plant J.* 2008;54:763–784.

Farnsworth NR, Akerele O, Bingel AS, Soejarto DD, Guo Z. Medicinal plant in therapy. *Bull World Health Organ.* 1985;63:965–981.

Frisvad JC, Smedsgaard J, Larsen TO, Samson RA. Mycotoxins, drugs and other extrolites produced by species in Penicillium subgenus Penicillium. *Stud Mycol.* 2004;49: 201–241.

Fulda S. Betulinic acid for cancer treatment and prevention. *Int J Mol Sci.* 2008;9(6): 1096–1107.

Gabor M, Eperjessy E. Antibacterial effect of fisetin and fisetinidin. *Nature.* 1966;212 (5067):1273. Doi:10.1038/2121273aO.

Gangadevi V, Muthumary J. Taxol, an anticancer drug produced by an endophytic fungus *Bartalinia robillardoides* Tassi, isolated from a medicinal plant, *Aegle marmelos* Correa e Roxb. *World J Microbiol Biotechnol.* 2008;24:717–724.

Geraets L, et al. Inhibition of LPS-induced pulmonary inflammation by specific flavonoids. *Biochem Biophys Res Commun.* 2009;382(3):598–603.

Giovannucci E, Ascherio A, Rimm EB, Stampfer MJ, Colditz GA, Willett WC. Intake of carotenoids and retinol in relation to risk of prostate cancer. *J Natl Cancer Inst.* 1995;87(23):1767–1776.

Giridharan P, Verekar SA, Khanna A, Mishra PD, Deshmukh SK. Anticancer activity of Sclerotiorin, isolated from an endophytic fungus *Cephalotheca faviolata*. *Indian J Expl Biol.* 2012;50(7):464–468.

Goldberg MR, et al. Phase I clinical study of the recombinant oncotoxin TP40 in superficial bladder cancer. *Clin Cancer Res.* 1995;1:57–61.

Govindachari TR, Viswanathan N. Alkaloids of *Mappia foetida*. *Phytochemistry.* 1972;11(12):3529–3531.

Gutheil WG, Reed G, Ray A, Dhar A. Crocetin: An agent derived from saffron for prevention and therapy for cancer. *Curr Pharm Biotechnol.* 2012;13(1):173–179.

Han GL, Li CM, Mazza G, Yang XG. Effect of anthocyanins rich fruit extract on PGE2 produced by endothelial cells. *J Hyg Res*. 2005;34:581–584.

Hartwell JL. *Plants Used Against Cancer: A Survey*. Lawrence, MA: Quarterman; 1982.

Honma Y, Ishii Y, Yamamoto-Yamaguchi Y, Sassa T, Asahi KI. Cotylenin A, a differentiation-inducing agent, and IFN-α cooperatively induce apoptosis and have an antitumor effect on human non-small cell lung carcinoma cells in nude mice. *Cancer Res*. 2003;63:3659–3666.

Hsiang YH, Herzberg R, Hecht S, Liu LF. Camptothecin induces protein-linked DNA breaks via mammalian DNA topoisomerase – I. *J Biol Chem*. 1985;260:14873–14878.

Hsiang CY, Ho TY, Lin CH, Wu K, Chang TJ. Analysis of upregulated cellular genes in pseudorabies virus infection: Use of mRNA differential display. *J Virol Methods*. 1996;62:11–19.

Itokawa H, Wang X, Lee K-H. Homoharringtonine and related compounds. In: Kingston DGI, Newman Cragg GM, eds. *Anticancer Agents from Natural Products*. Boca Raton, FL: Brunner-Routledge Psychology Press; 2005:47–70.

Jain KK. Use of bacteria as anticancer agents. *Exp Opin Biol Ther*. 2001;1(2):291–300.

Jemal A, Bray F, Center MM, Ferlay J, Ward E, Forman D. Global cancer statistics. *CA-Cancer J Clin*. 2011;61:69–90.

Jeong CH, et al. Gingerol suppresses colon cancer growth by targeting leukotriene A4 hydrolase. *Cancer Res*. 2009;69(13):5584–5591.

Jordan MA, Wilson L. Microtubules as a target for anticancer drugs. *Nat Rev Cancer*. 2004;4:253–265.

Joshi RD, Kulkarni NS. Screening of endophytic bacteria from anticancer medicinal plants. *IJSR*. 2015;4(5):1870–1873.

Jungman ML, Paulsen MT. The cyclin-dependent kinase inhibitor roscovitine inhibits RNA synthesis and triggers nuclear accumulation of p53 that is unmodified at Ser15 and Lys382. *Mol Pharmacol*. 2001;60:785–789.

Kantarjian HM, O'Brien S, Anderlini P, Talpaz M. Treatment of myelogenous leukemia: Current status and investigational options. *Blood*. 1996;87:3069–3081.

Karikas GA. Anticancer and chemopreventing natural products: Some biochemical and therapeutic aspects. *J B.U.ON*. 2010;15:527–638.

Kasukabe T, Okabe-Kado J, Honma Y. Cotylenin A, a new differentiation inducer, and rapamycin cooperatively inhibit growth of cancer cells through induction of cyclin G2. *Cancer Sci*. 2008;99(8):1963–1968.

Katiyar SK, Ahmad N, Mukhtar H. Green tea and skin. *Arch Dermatol*. 2000;136:989–994.

Kellard LR. Flavopiridol, the first cyclic-dependent kinase inhibitor to enter the clinic: Current status. *Expert Opin Investig Drugs*. 2000;9:2903–2911.

Khan N, Afaq F, Khusro FH, Adhami VM, Suh Y, Mukhthar H. Dual inhibition of PI3K/AKT and mTOR signaling in human non-small cell lung cancer cells by a dietary flavonoid fisetin. *Int J Cancer*. 2012;130(7):1695–1705.

Khanna V, Kannabiran K. Anticancer-cytotoxic activity of saponins isolated from the leaves of *Gymnema sylvestre* and *Eclipta prostrata* on HeLa cells. *Int J Green Pharm*. 2009;3:227–229.

Khanna GV, Kannabiran K. 3. Prediction of interaction of gymnemagenol and dasyscyphin c with cancer drug target proteins by in silico molecular docking studies. *Blue Biotech J*. 2013;1(2):221–230.

Kharwar RN, Verma VC, Strobel G, Ezra D. The endophytic fungal complex of *Catharanthus roseus* (L) G. Don. *Curr Sci*. 2008;95(2):228–233.

Kharwar RN, Mishra A, Gond SK, Stierle A, Stierle D. Anticancer compounds derived from fungal endophytes: Their importance and future challenges. *Nat Prod Rep (Royal Soc Chem)*. 2011;28:1208–1228.

Kim JM, Kim JS, Yoo H, Choung MG, Sung MK. Effects of black soybean [Glycine max (L.) Merr.] seed coats and its anthocyanidins on colonic inflammation and cell proliferation in vitro and in vivo. *J Agric Food Chem*. 2008;56(18):8427–8433.

Konuklugil B. Aryltetralin lignans from genus Linum. *Fitoterapia.* 1996;67:95–98.

Krohn K, et al. Massarilactones E-G, New metabolites from the endophytic fungus *Coniothyrium* sp., associated with the plant Artimisia maritime. *Chirality.* 2007;19(6):464–470.

Kumar AP, Graham H, Robson C, Thompson IM, Ghosh R. Natural products: Potential for developing phellodendron amurense bark extract for prostate cancer management. *Min Rev Med Chem.* 2010;10:388–397.

Laske DW, Youle RJ, Oldfield EH. Tumor regression with regional distribution of the targeted toxin TF-CRM107 in patients with malignant brain tumors. *Nat Med.* 1997;3:1362–1368.

Lee D, et al. Fungal metabolites, sorbicillinoid polyketides and their effects on the activation of peroxisome proliferator-activated receptor gamma. *J Antibiot.* 2005;58:615–620.

Lee HS, Seo EY, Kang NE, Kim WK. [6]-Gingerol inhibits metastasis of MDA MB-231 human breast cancer cells. *J Nutr Biochem.* 2008;(19):313–319.

Li JY, Strobel GA, Sidhu R, Hess WM, Ford E. Endophytic Taxol producing fungi from Bald Cypress *Taxodium distichum. Microbiology.* 1996;142(8):2223–2226.

Li JY, Sidhu RS, Ford E, Hess WM, Strobel GA. The induction of Taxol production in the endophytic fungus *Preiconia* sp. from *Torreya grandifolia. J Ind Microbiol.* 1998;20(5; May):259–264.

Ling-Hua M, Zhi-Yong L, Pommier Y. Non-camptothecin DNA topoisomerase I inhibitors in cancer therapy. *Curr Top Med Chem.* 2003;3(3; January):305–320.

Liu LF. DNA topoisomerase poisons as antitumor drugs. *Annu Rev Biochem.* 1989;58:351–375.

Lopez-Lazaro M, Willmore E, Austin CA. Cells lacking DNA topoisomerase II beta are resistant to genistein. *J Nat Prod.* 2007;70(5):763–767.

Magarvey NA, Keller JM, Bernan D, Dworkin M, Sherman DH. Isolation and characterization of novel marine-derived actinomycete taxa rich in bioactive metabolites. *Appl Environ Microbiol.* 2002;70:7520–7529.

Mahajan G, Balachandran L. Biodiversity in production of antibiotics and other bioactive compounds. *Adv Biochem Eng Biotechnol.* 2015;147:37–58.

Mahajan S, Bakshi S, Bansal D, Bhasin P. Isolation and characterisation of endophytes. *Int J Latest Res Sci Technol.* 2014;1(1):29–33.

Maheshwari R. What is an endophytic fungi? *Curr Sci.* 2006;90(10):1309.

Malik M, Zhao C, Schoene N, Guisti MM, Moyer MP, Magnuson BA. Anthocyanin rich extract from *Aronia meloncarpa* E. induces a cell cycle block in colon cancer but not normal colonic cells. *Nut Cancer.* 2003;46:186–196.

Mann J. Natural products in cancer chemotherapy: Past, present and future. *Nat Reve Cancer.* 2002;2(2):143–148.

Markovits J, et al. Inhibitory effects of the tyrosine kinase inhibitor genistein on mammalian DNA topoisomerase II. *Cancer Res.* 1989;49(18):5111–5117.

Mbaveng AT, et al. Evaluation of four Cameroonian medicinal plants for anticancer, antigonorrheal and anti-reverse transcriptase activities. *Environ Toxicol Pharm.* 2011;32:162–167.

Merlin NJ, Christudas NIVS, Kumar PP, Kumar M, Agastian P. Taxol production by endophytic *Fusarium solani* LCPANCF01 from *Tylophora indica. J Acad Indus Res.* 2012;1(5):281–285.

Mi Q, et al. Pervilleine A, a novel tropane alkaloid that reverses the multidrug-resistance phenotype. *Cancer Res.* 2001;61:842–850.

Mi Q, et al. Pervilleine F, a new tropane alkaloid aromatic ester that reverses the multidrug-resistance phenotype. *Anticancer Res.* 2003;23:3607–3616.

Minuzzo M, Marchini S, Broggini M, Faircloth G, D'incalci M, Mantovani R. Interference of transcriptional activation by the anti-neoplastic drug ET-743. *Proc Natl Acad Sci USA.* 2000;97:6780–6784.

Nam JY, Son K-H, Kim H-K, Han M-Y, Kim S-U, Choi J-D, Kwon B-M. Sclerotiorin and isocromophilone IV: Inhibitors of Grb2Shc interaction, isolated from Penicillium multicolor F1753. *J Microbiol Biotechnol.* 2000;10(4):544.

Newman DJ, Cragg GM. Natural products as sources of new drugs over the 30 years from 1981 to 2010. *J Nat Prod.* 2012;75(3):311–335.

Newman DJ, Cragg GM, Snader KM. Natural products as sources of new drugs over the period 1981–2002. *J Nat Prod.* 2003;66:1022–1037.

Nishikawa T, Nakajima T, Moriguchi M, Jo M, Sekoguchi S, Ishii M. A green tea polyphenol, epigallocatechin-3-gallate, induces apoptosis of human hepatocellular carcinoma, possibly through inhibition of Bcl-2 family proteins. *J Hepatol.* 2006;44:1074–1082.

Noble RL. The discovery of the vinca alkaloids—Chemotherapeutic agents against cancer. *Biochem Cell Biol.* 1990;68:1344–1351.

Nygren P, Larsson R. Overview of the clinical efficacy of investigational anticancer drugs. *J Int Med.* 2003;253:46–75.

Ohsumi K, et al. Novel combretastatin analogues effective against murine solid tumors: Design and structure-activity relationships. *J Med Chem.* 1998;41(16):3022–3032.

Oyagbemi AA, Saba AB, Azeez OI. Molecular targets of [6]-gingerol: Its potential roles in cancer chemoprevention. *Biofactors.* 2010;36(3):169–178.

Pairet L, et al. Azaphilones with endothelin receptor binding activity produced by *Penicillium sclerotiorum*: Taxonomy, fermentation, isolation, structure elucidation and biological activity. *J Antibiot.* 1995;48(9; September):913–923.

Pastan I, Kreitman RJ. Immunotoxins for targeted cancer therapy. *Adv Drug Deliv Rev.* 1998;31:53–88.

Pettit GR, Singh SB, Niven ML, Hamel E, Schmidt JM. Isolation, structure, and synthesis of combretastatins A-1 and B-1, potent new inhibitors of microtubule assembly, derived from *Combretum caffrum. J Nat Prod.* 1987;50(1):119–120.

Pommier Y. Topoisomerase I inhibitors: Camptothecins and beyond. *Nat Rev Cancer.* 2006;6(10; October):789–802.

Pomponi AS. The bioprocess-technological potential of the sea. *J Biotechnol.* 1999;70:513.

Potmeisel M, Pinedo H. *Camptothecins: New Anticancer Agents.* Boca Raton, FL: CRC Press; 1995.

Powell RG, Weisleder D, Smith Jr CR, Rohwedder WK. Structures of harringtonine, isoharringtonine, and homoharringtonine. *Tetrahedron Lett.* 1970;11:815–818.

Puri RK. Development of a recombinant interleukin-4 *Pseudomonas* exotoxin for therapy of glioblastoma. *Toxicol Pathol.* 1999;27:53–57.

Ravindran J, Prasad S, Aggarwal BB. Curcumin and cancer cells: How many ways can curry kill tumor cells selectively? *Aaps J.* 2009;11(3):495–510.

Ravindra NK, Ashish M, Surendra KG, Andrea S, Donald S. Anticancer compounds derived from fungal endophytes: Their importance and future challenges. *Nat Prod Rep.* 2011(28; April):1208–1228.

Rinehart KL. Antitumor compounds from tunicates. *Med Res Rev.* 2000;20:1–27.

Rocha AB da, Lopez RM, Schwartsmann G. Natural products in anticancer therapy. *Curr Opin Pharmacol.* 2001;1:364–369.

Ryan RP, Germaine K, Franks A, Ryan DJ, Dowling DN. Bacterial endophytes: Recent developments and applications. *FEMS Microbiol Lett.* 2008;278:1–9.

Saeidnia S, Abdollahi M. Perspective studies on novel anticancer drugs from natural origin: A comprehensive review. *Int J Pharmacol.* 2014;10(2):90–108.

Sanjana K, Suruchi G, Maroof A, Manoj KD. Endophytic fungi from medicinal plants: A treasure hunt for bioactive metabolites. *Phytochem Rev.* 2012;11(4):487–515.

Scherlach K, Boetteger D, Remme N, Hearweck C. The chemistry and biology of cytochalasans. *Nat Prod Rep.* 2010;27:869–886.

Schiff PB. Promotion of microtubule assembly in vitro by Taxol. *Nature.* 1979;277:665–667.

Schiff PB, Horowitz SB. Taxol stabilizes microtubules in mouse fibroblast cells. *Proc Natl Acad Sci.* 1980;77(3; March):1561–1565.

Schmeller T, Latz-Brüning B, Wink M. Biochemical activities of berberine, palmatine and sanguinarine mediating chemical. *Phytochemistry*. 1997;44(2; January):257–266.

Schweitzer G, et al. Summary of the workshop on drug development, biological diversity, and economic growth. *J Natl Cancer Inst*. 1991;83:1294–1298.

Senft C, et al. The nontoxic natural compound Curcumin exerts anti-proliferative, anti-migratory, and anti-invasive properties against malignant gliomas. *BMC Cancer*. 2010;10:491.

Shukla S, Mehta A. Anticancer potential of medicinal plants and their phytochemicals: A review. *Braz J Bot*. 2015;38(2):199–210.

Shweta S, et al. Isolation of endophytic bacteria producing the anti-cancer alkaloid campto-thecine from *Miquelia dentata* Bedd. (Icacinaceae). *Phytomedicine*. 2013;20:913–917.

Strobel GA. Microbial gifts from rain forests. *Can J Plant Pathol*. 2002;24(1):14–20.

Strobel G, Yang X, Sears J, Kramer R, Sidhu RS, Hess WM. Taxol from Pestalotiopsis microspora, an endophytic fungus of *Taxus wallichiana*. *Microbiology*. 1996;142(2; February):435–440.

Suffness M. *Taxol, Science and Applications*. New York, NY: CRC Press; 1995.

Sun Y, Xun K, Wang Y, Chen X. A systematic review of the anticancer properties of berber-ine, a natural product from Chinese. *Anti-Cancer Drugs*. 2009;20(9):757–769.

Takahashi M, Kawamura A, Kato N, Nishi T, Hamachi I, Ohkanda J. Phosphopeptide-dependent labeling of 14-3-3 ζ proteins by fusicoccin-based fluorescent probes. *Angew Chem, Int Ed*. 2012;51:509.

Takaku M, et al. Halenaquinone, a chemical compound that specifically inhibits the second-ary DNA binding of RAD51. *Genes Cells*. 2011;16(4):427–436.

Taylor JR, Wilt VM. Probable antagonism of warfarin by green tea. *Ann Parmacother*. 1999;33:426–428.

Usui T, Kondoh M, Cui CB, Mayumi T, Osada H. Tryprostatin A, a specific and novel inhibi-tor of microtubule assembly. *Biochem J*. 1998;333:543–548.

Verekar SA, et al. Anticancer activity of new depsipeptide compound isolated from an endo-phytic fungus. *J Antibiot*. 2014;67:697–701.

Wall ME, Wani MC, Cook CE, Palmler KH, McPhail AT. Plant antitumour agents I. The isolation and structure of camptothecin, a novel alkaloidal leukemia and tumour inhibi-tor from *Camptotheca acuminata*. *J Am Chem Soc*. 1966;88(16; August):3888–3890.

Wani MC, Taylor HL, Wall ME, Coggon P, McPhail AT. Plant antitumor agents. VI. The isolation and structure of Taxol, a novel antileukemic and antitumor agent from *Taxus brevifolia*. *J Am Chem Soc*. 1971;93(9):2325–2327.

Weiss SG, et al. Potential anticancer agents II. Antitumour and cytotoxic lignans from *Linum album* (Linaceae). *J Pharm Sci*. 1975;64:95–98.

Yamamoto K, Takase H, Abe K, Saito Y, Suzuki A. Pharmacological studies on antidiarrheal effects of a preparation containing Berberine and Geranii Herba. *Nippon Yakurigaku Zasshi*. 1993;101(3; March):169–175.

Yang T, et al. The novel agent ophiobolin O induces apoptosis and cell cycle arrest of MCF-7 cells through activation of MAPK signaling pathways. *Bioorg Med Chem Lett*. 2012;22(1):579–585.

Young DH, Michelotti EJ, Sivendell CS, Krauss NE. Antifungal properties of Taxol and vari-ous analogues. *Experientia*. 1992;48(9; September):882–885.

Zhang M, et al. Cloning and expression of the gene coding for IL-2(60)-PE40 a molecular targeted protein. *Chin Med Sci J*. 1995;10:136–140.

Zhou DC, Zittoun R, Marie JP. Homoharringtonine: An effective new natural product in cancer chemotherapy. *Bull Cancer*. 1995;82:987–995.

4 Treatment of Cancer

Nathiya Shanmugam

CONTENTS

INTRODUCTION

Cancer is the second leading cause of death globally. Early diagnosis and treatment have resulted in significant improvement in survival. The treatment of cancer depends on age, health status, and type, size, and stage of the cancer.

The different types of cancer treatment include:

1. Surgery
2. Radiation therapy

3. Immunotherapy
4. Targeted therapy
5. Stem cell therapy
6. Hormone therapy
7. Chemotherapy

SURGERY (AMERICAN CANCER SOCIETY)

Surgery is the oldest treatment. If the cancer has not metastasized to another site, it can be removed by surgery. Surgery is often seen in the removal of testicle, breast, and prostate cancer. The goal of the surgery is to remove the tumor or the entire organ (Subotic et al. 2012). It is often essential in removing primary tumor and also in determining the extent of disease and to control symptoms. Conventional surgery makes large incisions in the skin and other layers of the body. Hence, newer surgical techniques are now available that use different instruments and are less painful for the patient and require shorter recovery times. Newer techniques are discussed in the following sections.

LASER SURGERY

Laser treatment is carried out on an outpatient basis, and it is carried out without anesthesia. It is a noncontact technique with less pain and bleeding than surgery (Orihuela and Smith 1989). Lasers are also used in photoablation or photocoagulation surgery to destroy tissues or seal tissues or blood vessels. This type of surgery is often used to relieve symptoms, such as when large tumors block the windpipe (trachea) or swallowing tube (esophagus), causing problems with breathing or eating (American Cancer Society 2016). The main types of laser used in the surgery are carbon dioxide, argon, and neodymium:yttrium-aluminum-garnet.

CRYOSURGERY

Cryosurgery uses a liquid nitrogen or argon gas to produce extreme cold to destroy abnormal cells. A cryoprobe is used to circulate the liquid nitrogen or argon gas, which is placed in contact with the tumor. The ice crystal balls are produced around the probe and freeze the nearby tissues. Ultrasound or a computed tomography scan is used to monitor the freezing of the cells, limiting damage to nearby tissue. It is used to treat numerous cancers, including breast, skin, cervix, AIDS-related Kaposi sarcoma, prostate, and bone cancers (National Cancer Institute 2015).

ELECTROSURGERY

High-frequency electrical current is used to kill cancer cells in electrosurgery. It is mainly used in treating skin cancer (Sheridan and Dawber 2000).

RADIOFREQUENCY ABLATION

This is a percutaneous ablation technique that is the most commonly used technique because of its safety, efficacy, and ease of use. It is used in the treatment of primary, recurrent, and metastatic solid tumors involving lungs, liver, kidney, adrenal gland, bone, and the musculoskeletal system. Radiofrequency ablation involves a flow of electric alternating current through tissue; thereby, ionic agitation and resistive heating of tissue occur. Deposition of thermal energy into the tumors results in thermal injury having a tumoricidal effect. It can be done on an outpatient basis and requires only a minimal hospital stay (Tatli et al. 2012).

MOHS SURGERY

Known also as Mohs micrographic surgery, Mohs surgery is mainly used in skin cancers. The dermatologist performs both surgical excision of skin cancer and microscopic examination of the surgical margin to confirm that all the skin cancer cells are removed.

LAPAROSCOPIC SURGERY

This is also known as minimally invasive surgery because it involves a smaller incision (0.5–1.5 cm), with less intraoperative blood loss and postoperative pain, and a short hospital stay (Angst et al. 2010). It is mainly used for endometrial (Hauspy et al. 2010), colon (Kahnamoui et al. 2007), and rectal cancers.

THORACOSCOPIC SURGERY

In this type of surgery, a thin scope with a video camera can be put into the chest through a small incision. Small tumors on the surface of the lung can be removed, fluid can be drained, and small tissues on the lining of the chest wall can be taken out. In a non-small cell lung cancer, video-assisted thoracoscopic surgery is used, which appears to lower morbidity and improve survival status (Whitson et al. 2008).

ROBOTIC SURGERY

One of the advanced methods in noninvasive surgery is robotic technology. It is a type of laparoscopic (or thoracoscopic) surgery where the doctor sits at a control panel and uses precise robotic arms to control the scope and other special instruments. Robotic surgery is sometimes used to treat cancers of the colon, prostate, and uterus. The advantages of this type of surgery are largely the same as laparoscopic and thoracoscopic surgery: it can help reduce blood loss during surgery and pain afterward. It can also shorten hospital stays and allow people to heal faster. Some of the robots' capabilities in general for surgeons include improved visualization, the ability to manage multiple tasks simultaneously, stability and greater accuracy, optimization for the particular environment, improvement in the surgeon's skills by reducing both performance time and errors, enhanced dexterity by tremor abolition, motion scaling and reduced ergonomic problems of surgeons in longer procedures as surgeons can sit down and use their hands and fingers through ergonomically designed controls, high-quality 3D vision, facilitation of complex procedures, enhancement of dexterity to facilitate microscale operations, and development of virtual simulator trainers to enhance the ability to learn new complex operations (Mohammadzadeh and Safdari 2014).

RADIATION THERAPY

Radiation therapy uses high-energy radiation to shrink and kill cancer cells. X-rays and gamma rays are used for cancer treatment. There are two types of radiotherapy: external and internal. In external therapy, the radiation is delivered by a machine outside the body into the tumor. The machine delivers the high-energy x-rays. Internal radiation therapy is also known as brachytherapy; in this the radioactive material is placed near the cancer cells. This is used particularly in the routine treatment of gynecological and prostate malignancies as well as in situations where retreatment is indicated, based on its short-range effects. Systemic radiation therapy uses radioactive iodine that circulates in the blood and kills the cancer cells (Jackson and Bartek 2009).

Radiation therapy directly damages the DNA of the cancer cells and kills it, otherwise it produces free radicals inside the cell, which in turn kill the cancer cells. Radiation can damage both normal and cancer cells. Normal cells usually repair themselves and restore their functions at a faster rate than the cancer cells. Radiation therapy can be employed in cure as well as in palliative treatment to relieve the symptoms of cancer. Further, it can be used in combination with other therapies (Begg et al. 2011). It is used before surgery to shrink the tumor and after surgery to destroy the microscopic tumor cells that may be left behind after surgery (Baskar et al. 2012).

Early cancers curable with radiation therapy alone include skin, lung, cervix, and head and neck cancers. Cancers curable with radiation therapy in combination with other modalities are breast, rectal, anal, advanced lymphomas, endometrial, and bladder carcinoma (Baskar et al. 2012).

IMMUNOTHERAPY

The immune system will detect and destroy abnormal cells, thereby preventing the development of cancers. But in some cases, a cancer cell avoids detection and destruction by the immune system. In the field of cancer, immunology developed a new method called *immunotherapy* to treat cancer. Immunotherapy increases the strength of the immune system against a tumor either by stimulating specific components in the immune system or by suppressing the signals produced by the cancer cell which suppress the immune system.

IMMUNE CHECKPOINT MODULATORS

Immunotherapy blocks certain proteins called *immune checkpoint proteins* to limit the strength and duration of immune responses. Blocking the activity of the proteins will release the brake on the immune system and increase the strength of the immune system to destroy the cancer cells. Cytotoxic T lymphocyte-associated-protein 4 (CTLA-4) and programmed cell death-protein 1 (PD-1) are negative regulators of T-cell activation and can contribute to immune evasion by tumor cells. Monoclonal antibody inhibitors of these checkpoint pathways enhance T-cell proliferation and function, perpetuating T-cell activation and reawakening the silenced antitumor immune response.

The first drug to be approved by the U.S. Food and Drug Administration (FDA) was ipilimumab to treat advanced melanoma. It blocks the activity of a checkpoint protein known as CTLA4, expressed on the cytotoxic T lymphocytes. CTLA4 acts as a "switch" to inactivate these T cells, thereby reducing the strength of immune responses; ipilimumab binds to CTLA4 and prevents it from sending its inhibitory signal. Ipilimumab is limited by its adverse effects, like immune-related adverse effects such as pruritus, rash, diarrhea, colitis, hypophysitis, and increases in compounds detected by liver function tests (Eggermont et al. 2015; Schadendorf et al. 2015).

PD-1 CHECKPOINT INHIBITORS

Pembrolizumab and nivolumab are the programmed cell death-protein 1 (PD-1) checkpoint inhibitors that have demonstrated significant efficacy in patients with ipilimumab-naïve and ipilimumab-pretreated melanoma. Both of these drugs were approved in 2015 for the treatment of patients with unresectable or metastatic disease and increase overall survival (Redman et al. 2016). In patients with advanced melanoma, pembrolizumab and nivolumab show superior efficacy compared to ipilimumab in overall response rate and survival. In research studies, both proved to be even more effective than ipilimumab, while causing fewer adverse effects.

The PD-1 inhibitor nivolumab, pembrolizumab, and the PD-L1 (programmed cell death ligand 1 [PD-L1]) inhibitor atezolizumab show significant improvement in patients with advanced squamous non-small cell lung cancer (NSCLC) with disease progression and reducing the risk of death (Brahmer et al. 2015; Hui et al. 2016; Smith et al. 2016). Pembrolizumab and the novel PD-L1 inhibitors avelumab and durvalumab have also demonstrated efficacy as monotherapy in Phase I/II trials in advanced urothelial cancer (Plimack et al. 2015; Apolo et al. 2016; Massard et al. 2016). The combination of nivolumab and ipilimumab is also under investigation in advanced renal cell carcinoma and has demonstrated preliminary efficacy in a Phase I trial (Bristol-Myers Squibb 2015; Hammers et al. 2015).

VIROTHERAPY

Virotherapy effectively uses oncotropic and oncolytic viruses that have the ability to find and destroy cancerous cells in the body.

Adapted ECHO-7 virus Rigvir immunotherapy (oncolytic virotherapy) prolongs survival in patients with melanoma after surgical excision of the tumor (Doniņa et al. 2015).

TALIMOGENE LAHERPAREPVEC (T-VEC)

This is an attenuated oncolytic virus that expresses granulocyte macrophage colony stimulation factor (GM-CSF), which stimulates the tumor-specific T-cell response. The oncolytic intralesional therapy Talimogene laherparepvec (T-VEC) was approved for treating unresectable melanoma. In ongoing clinical trials, T-VEC are being investigated in combination with immunotherapies like checkpoint inhibitors (Long et al. 2015; Puzanov et al. 2015).

DARATUMUMAB

This targets CD38 and acts by two mechanisms: it destroys cancer cells directly or enhances the immune system to fight against cancer. The FDA approved daratumumab in the treatment of multiple myeloma.

TARGETED THERAPY

This therapy is designed to target the specific molecules that help cancer cell growth and progression. It is essential to specifically identify and quantify protein targets in tumor tissues for the reasonable use of such targeted therapies. In tumors the signaling pathways' deregulation leads to cancer cell growth and proliferation. These signaling pathways become the focus of the development of targeted cancer therapies. The deregulation of downstream signaling molecules might also play an important role in the success of such therapeutic approaches.

For these reasons, the analysis of tumor-specific protein expression profiles prior to therapy has been suggested as the most effective way to predict possible therapeutic results. Reverse phase protein microarray (RPPA) is a promising technology that meets these requirements while enabling the quantitative measurement of proteins. Together with recently developed protocols for the extraction of proteins from formalin-fixed, paraffin-embedded (FFPE) tissues, RPPA may provide the means to quantify therapeutic targets and diagnostic markers in the near future and reliably screen for new protein targets. The communication in the protein network is responsible for activation and deactivation of involved proteins, which is often altered in cancer due to aberrant cellular functions resulting in proliferation, apoptosis, differentiation, survival, invasion, and metastasis. RPPA analyzes whole protein signaling networks and provides fundamental information about the functional state of signaling pathways.

Human epidermal growth factor receptors 1 (EGFR) and 2 (HER2) are the kinase receptors. Both kinases act as targets for anticancer drugs and can be analyzed by clinically approved tests, such as immunohistochemistry (IHC) and fluorescence in situ hybridization (FISH). EGFR overexpression is often found in human cancers; in gliomas, this deregulation is associated with structural rearrangements leading to in-frame deletions in the extracellular domain of the receptor (Hynes and Lane 2005). Breast cancers can be mediated either by transcriptional activation or gene amplification (Emens 2005). The HER2 status of breast cancer patients not only has a predictive value, but the receptor itself is also a target for the monoclonal anti-HER2 antibody trastuzumab (Table 4.1) (Piccart et al. 2005).

STEM CELL THERAPY

High doses of chemotherapy or radiotherapy kill the cancer cells that divide rapidly. High-dose treatment also destroys the bone marrow because it also divides rapidly.

TABLE 4.1
Drugs Currently Used for Targeted Therapy

Drugs	Target	Tumor Type
Trastuzumab	HER2	Metastatic breast cancer, gastric cancer
Cetuximab	EGFR	Metastatic colorectal cancer
Bevacizumab	VEGF	Colorectal cancer
Lapatinib	EGFR, HER2	Breast cancer
Sunitinib	VEGFR, PDGFR, cKit, Flt-3	Renal cell cancer
Rapamycin RAD001	mTOR	Breast, prostate, renal cancer

Source: Malinowsky K, et al. *Journal of Cancer.* 2011;2:26–35.

Bone marrow transplantation (BMT) and peripheral blood stem cell transplantation (PBSCT) are procedures that restore stem cells that have been destroyed by high doses of chemotherapy or radiation therapy. A stem cell transplant can be used to infuse healthy stem cells into the body to stimulate new bone marrow growth, suppress the disease, and reduce the possibility of a relapse. Stem cells can be found in the bone marrow, circulating blood (peripheral blood stem cells), and umbilical cord blood.

There are three different types of stem cell transplants:

1. Autologous (using one's own stem cells)
2. Allogenic (using someone else's stem cells)
3. Syngeneic (stem cells from an identical twin)

Stem cells usually recover the ability to produce stem cells in the body after chemotherapy or radiotherapy. But in multiple myeloma and some types of leukemia, a stem cell transplant may work against cancer directly. This happens because of an effect called graft-versus-tumor that can occur after allogenic transplants. Graft-versus-tumor occurs when white blood cells from your donor (the graft) attack any cancer cells that remain in your body (the tumor) after high-dose treatments. This effect improves the success of the treatments. Stem cell transplants are most often used in leukemia, lymphoma, neuroblastoma, and multiple myeloma.

Side effects include nausea, vomiting, fatigue, loss of appetite, mouth sores, and hair loss. Potential long-term risks include complications of the pretransplant chemotherapy and radiation therapy, such as infertility, cataracts, secondary cancers, and damage to the liver, kidneys, lungs, and heart.

HORMONE THERAPY FOR CANCER

In cancer, hormone therapy slows or inhibits the growth of cancers that utilize hormones for their growth. Hormone therapy is used to treat breast or prostate cancer and also reduces the symptoms of cancer in men with prostate cancer who are not undergoing surgery or radiation therapy. When used with other cancer therapy, it has been shown to reduce the size of the tumor before surgery or radiation therapy. It is called *neoadjuvant therapy*.

Side effects mainly depend on the type of hormone therapy. Men receiving hormone therapy for prostate cancer have hot flashes, libido changes, bone weakness, diarrhea, and nausea.

For women receiving hormone therapy for breast cancer, side effects include hot flashes, vaginal dryness, irregular menstruation, nausea, and mood changes (National Cancer Institute 2015).

CHEMOTHERAPY

In the present methods of treatment about one-third of patients are cured with local treatment strategies like surgery or radiotherapy, if the tumor is localized at the time of diagnosis. Early diagnosis increases the cure rate with the local treatment, and in the remaining cases, a systemic approach with chemotherapy is required for effective cancer management (Table 4.2). In locally advanced disease, chemotherapy is combined with radiotherapy for surgical resection, and these combined treatment modalities improve the clinical outcomes.

Chemotherapy is used in three main clinical situations:

1. Primary treatment in patients with advanced cancer for which alternate treatment does not exist (e.g., Hodgkin's and non-Hodgkin's lymphomas, acute myelogenous leukemia, and germ cell cancer)
2. Neoadjuvant chemotherapy in patients with localized cancer for which alternative therapy is available but which is less than completely effective (e.g., anal cancer, breast cancer, esophageal cancer, and laryngeal cancer)
3. As an adjuvant to local treatments such as surgery and radiation therapy— termed *adjuvant chemotherapy*—to reduce the incidence of local and systemic recurrence and to improve the overall survival of the patients (e.g., colon cancer and gastric cancer)

CLINICAL PHARMACOLOGY OF CANCER CHEMOTHERAPEUTIC DRUGS

Knowledge of tumor cell proliferation and an understanding of the mechanisms of action of cancer chemotherapeutic agents are important in selecting a suitable regimen for the patient with cancer.

Leukemias

Acute Leukemia

Acute Lymphoblastic Leukemia   Acute lymphoblastic leukemia (ALL) is most common in childhood. Methotrexate, corticosteroids, 6-mercaptopurine, cyclophosphamide, vincristine, and daunorubicin have been used against this disease.

Acute Myelogenous Leukemia   Acute myelogenous leukemia (AML) is the most common leukemia in adults. The effective agent for leukemia is cytarabine; however, it is best used in combination with anthracycline, Idarubicin.

TABLE 4.2
Various Types of Anticancer Agents with Their Mechanisms and Clinical Uses

Class	Subclass	Mechanism of Action	Clinical Uses
Antimetabolites	Folate antagonists: • Methotrexate • Pemetrexed • Pralatrexate	Inhibits the dihydrofolate reductase and thus affects nucleoside metabolism	Acute lymphoblastic leukemia (ALL), choriocarcinoma, breast cancer, neck and head cancers, lung cancer, cervical cancer
	Pyrimidine antagonists: • 5-Flourouracil • Cytarabine • Gemcitabine • Capecitabine	Block pyrimidine nucleotide formation by incorporation into newly synthesized DNA	Breast cancer, colorectal cancer, gastroesophageal cancer, head and neck cancer, hepatocellular cancer, pancreatic cancer, cancers of prostate and bladder
	Purine antagonists: • 6-Mercaptopurine • 6-Thioguanine	Act as fraud substrate for biochemical reactions and inhibit the synthetic steps during S-phase of replication	Acute and chronic myelogenous leukemia, acute lymphocytic leukemia, acute myelomonocytic leukemia
Genotoxic agents (bind to DNA and directly/indirectly affect the replication, which induces the apoptosis)	Alkylating agents: • Mechlorethamine • Chlorambucil • Bendamustine	Forms DNA cross-links, resulting in inhibition of DNA synthesis and function	Hodgkin's and non-Hodgkin's lymphomas
	Cyclophosphamide		Breast cancer, ovarian cancer, non-Hodgkin's lymphoma, chronic lymphoblastic leukemia, soft tissue sarcoma, neuroblastoma, Wilms' tumor, rhabdomyosarcoma
	Melphalan		Multiple myeloma, breast and ovarian cancers
	Thiotepa	Forms DNA cross-links, resulting in inhibition of DNA synthesis and function	Breast cancer, ovarian cancer, superficial bladder cancer
	Busulfan		Chronic myelogenous leukemia
	Carmustine Lomustine		Brain cancer, Hodgkin's and non-Hodgkin's lymphomas, brain cancer

(Continued)

TABLE 4.2 (*Continued*)
Various Types of Anticancer Agents with Their Mechanisms and Clinical Uses

Class	Subclass	Mechanism of Action	Clinical Uses
	Temozolomide	Methylates DNA and inhibits DNA synthesis and function	Brain cancer, melanoma
	Procarbazine		Hodgkin's and non-Hodgkin's lymphomas, brain
	Dacarbazine		tumors, melanoma, soft tissue sarcoma
Platinum compounds	Cisplatin	Forms intrastrand and interstrand DNA cross-links; binding to nuclear and cytoplasmic proteins	Non-small cell and small cell lung cancer, breast cancer, bladder cancer, cholangiocarcinoma, gastroesophageal cancer, head and neck cancer, ovarian cancer, germ cell cancer
	Carboplatin		Non-small cell and small cell lung cancer, breast cancer, bladder cancer, head and neck cancer, ovarian cancer
	Oxaliplatin		Colorectal cancer, gastroesophageal cancer, pancreatic cancer
	Anthracyclines: • Epirubicin • Doxorubicin • Dactinomycin	Oxygen free radicals bind to DNA causing single- and double-strand DNA breaks; inhibits topoisomerase II; intercalates into DNA	Breast cancer, acute leukemia, endometrial cancer, thyroid cancer, Wilms' tumor, Ewing's sarcoma, rhabdomyosarcoma, neuroblastoma.
Antitumor antibiotics	Enzyme inhibitors: • Etoposide • Topotecan • Irinotecan • Idarubicin	Etoposide: Inhibits topoisomerase II thus prevents resealing of DNA, which leads to cell death	Non-small cell and small cell lung cancer; non-Hodgkin's lymphoma, gastric cancer
		Topotecan/Irinotecan: Inhibits topoisomerase I, which allows single strands to break in DNA but not affect resealing	Colorectal cancer, gastroesophageal cancer, non-small cell and small cell lung cancer, ovarian cancer
			Acute myeloid leukemia

(Continued)

TABLE 4.2 (*Continued*)
Various Types of Anticancer Agents with Their Mechanisms and Clinical Uses

Class	Subclass	Mechanism of Action	Clinical Uses
Mitotic spindle inhibitors	Vinca alkaloids: • Vincristine • Vinblastine • Vinorelbine	Arrest the cell division in metaphase by binding to tubulin	Vincristine: Acute lymphocytic leukemia, Wilms' tumor, rhabdomyosarcoma, breast-cervical-ovarian cancer Vinblastine: Testicular carcinoma, Hodgkin's disease, Kaposi's sarcoma, cancer of breast, lungs, bladder
	Taxanes derivatives: • Paclitaxel • Docetaxel • Cabazitaxel	Stabilize polymerization of tubulins and inhibit the disassembly of microtubules	Cancer of breast, ovary, lungs, head and neck
Newer agents	Protein tyrosine kinase inhibitors: • Imatinib	By inhibiting this enzyme, inhibit proliferation of myeloid cell	Chronic myeloid leukemia, gastrointestinal stromal cell tumor
	Growth factor receptor (EGFR) inhibitors: • Gefitinib • Erlotinib	Activation of EGFR induces dimerization and intracellular activation of protein tyrosine kinase	Metastatic non-small cell lung cancer, solid tumors
	Proteosome inhibitors: • Bortezomib	Prevents degradation of intracellular protein leading to activation of signaling cascade, cell cycle arrest, and apoptosis	Refractory and relapsed multiple myeloma
	Monoclonal antibodies: • Rituximab • Alemtuzumab • Trastuzumab	Agents target CD20, CD52, B-cell antigen, which activates apoptosis; trastuzumab targets against human epidermal growth factor receptor protein-2 (HER-2) and induces the apoptosis in breast cancer	B-cell lymphoma, chronic lymphocytic leukemia, T-cell lymphoma, and breast cancer
	Aromatase inhibitors: • Anastrozole • Letrozole • Exemestane		Estrogen receptor (ER) positive metastatic breast cancer in postmenopausal women who are resistant to tamoxifen therapy

Chronic Myelogenous Leukemia   The tyrosine kinase inhibitor imatinib is considered as standard first-line therapy in previously untreated patients with chronic myelogenous leukemia. Dasatinib and nilotinib were used in Imatinib-resistant cases.

Chronic Lymphocytic Leukemia   Chlorambucil and cyclophosphamide are the two most widely used alkylating agents for this disease. In most cases cyclophosphamide is combined with vincristine and prednisone or with doxorubicin. Bendamustine is the newest alkylating agent used, either as monotherapy or in combination with prednisone. The purine nucleoside analog fludarabine given alone, in combination with cyclophosphamide, or combined with the anti-CD20 antibody rituximab is also effective. Monoclonal antibody-targeted therapies are being widely used in chronic lymphocytic leukemia, especially in relapsed or refractory disease. Rituximab is an anti-CD20 antibody that can be used in these cases.

Hodgkin's and Non-Hodgkin's Lymphomas

Hodgkin's Lymphoma

More recently, the anthracycline-containing regimen termed ABVD (doxorubicin, bleomycin, vinblastine, and dacarbazine) has been shown to be more effective and less toxic. An alternative regimen utilizes a 12-week course of combination chemotherapy (doxorubicin, vinblastine, mechlorethamine, vincristine, bleomycin, etoposide, and prednisone), followed by involved radiation therapy.

Non-Hodgkin's Lymphoma

Initial therapy with anthracycline-containing regimen CHOP (cyclophosphamide, doxorubicin, vincristine, and prednisone) has been considered to be the best treatment. In a randomized Phase III clinical trial, the combination of CHOP with rituximab improved response rates, disease-free survival, and overall survival compared with CHOP chemotherapy alone.

Multiple Myeloma

Recently, the combination of lenalidomide and dexamethasone or the proteosome inhibitor bortezomib and melphalan and prednisone have been shown to be more effective as first-line therapy. Thalidomide has been used in combination with dexamethasone, which will increase the response rate. Bortezomib was the first drug approved as the first-line treatment for relapsing or refractory cases of multiple myeloma.

Breast Cancer

Stage I and Stage II disease

Women with small primary tumors and negative axillary lymph node dissections (Stage I) are treated with surgery alone. Women with node-positive disease have a high risk of recurrence. Postoperative use of systemic adjuvant chemotherapy with six cycles of cyclophosphamide, methotrexate, and fluorouracil (CMF protocol) or of fluorouracil, doxorubicin, and cyclophosphamide (FAC) has been shown to significantly reduce the relapse rate and prolong survival. Alternative regimens include four cycles of doxorubicin and cyclophosphamide and six cycles of fluorouracil, epirubicin, and cyclophosphamide (FEC). Each of these chemotherapy regimens has benefited women with Stage II breast cancer. Tamoxifen is beneficial in postmenopausal women when used alone or in combination with cytotoxic chemotherapy.

Stage III and Stage IV Disease

For advanced cancer, the current treatment options are only palliative. Combination chemotherapy, endocrine therapy, or both increase the response rate. In hormone receptor–positive tumors, the aromatase inhibitors anastrozole and letrozole are now approved as first-line drugs. The aromatase inhibitors and exemestane are approved as second-line therapy following treatment with tamoxifen. The anthracyclines, the taxanes, along with the microtubule inhibitor have a broad range of activity against this disease.

Prostate Cancer

Prostate cancer was the second cancer shown to be responsive to hormonal manipulation. Luteinizing hormone-releasing hormone (LHRH) agonists—including leuprolide and goserelin agonists, alone or in combination with an anti-androgen (e.g., flutamide, bicalutamide, or nilutamide)—are the preferred approach. In hormone-refractory prostate cancer, mitoxantrone and prednisone are approved. Estramustine, an antimicrotubule agent, or a combination with either etoposide or a taxane such as docetaxel or paclitaxel increases the response rates.

Colorectal Cancer

For colorectal cancer, the following treatment is used: four active cytotoxic agents include 5-FU, the oral fluoropyrimidine capecitabine, oxaliplatin, and irinotecan; and an active biologic agent like the anti-VEGF antibody bevacizumab and the anti-EGFR antibodies cetuximab and panitumumab.

Gastroesophageal Cancers

Chemotherapy, using either intravenous 5-FU or oral capecitabine, is used to target gastroesophageal cancers. In addition, cisplatin-based regimens in combination with either irinotecan or one of the taxanes (paclitaxel or docetaxel) or with biologic agent trastuzumab provide significant clinical benefit in gastric cancer patients whose tumors overexpress the HER-2/neu receptor. In addition, neoadjuvant approaches with combination chemotherapy and radiation therapy prior to surgery appear to have promise in select patients.

Pancreatic Cancer

Gemcitabine is used alone for metastatic pancreatic cancer; however, in combination with erlotinib, which targets the EGFR-associated tyrosine kinase in advanced or metastatic pancreatic cancer, it improves the response rate. The use of adjuvant chemotherapy with either single-agent gemcitabine or 5-FU/leucovorin in patients with early-stage pancreatic cancer who have undergone successful surgical resection also shows improvement.

Lung Cancer

Two subtypes of lung cancer include non-small cell lung cancer and small cell lung cancer. The small cell lung cancer is the most aggressive form of lung cancer and is initially treated with a platinum-based combination regimen that includes cisplatin and irinotecan. Patients who fail the platinum-based therapy can be treated with topoisomerase I inhibitor topotecan. With early diagnosis, it may be curable using the combination of chemotherapy and radiation therapy.

In non-small cell lung cancer diagnosed at early stages, surgical resection can be performed for patient cure, and platinum-based chemotherapy provides survival benefits. Radiation therapy can be used in distant metastasis for palliation of pain, airway obstruction, or bleeding, and to treat patients whose performance status would not allow for more aggressive treatments. In advanced stage, a platinum-based compound cisplatin or carboplatin is recommended. Paclitaxel, vinorelbine, cetuximab, and bevacizumab can also be used. Finally, first-line therapy with an EGFR tyrosine kinase inhibitor, such as erlotinib or gefitinib, significantly improves outcomes in NSCLC patients with sensitizing EGFR mutations.

Ovarian Cancer

Ovarian cancer remains occult and symptomatic only when it has already metastasized to the peritoneal cavity. For Stage I disease, abdominal radiotherapy offers benefits along with cisplatin and cyclophosphamide. In Stages III and IV, carboplatin plus paclitaxel becomes the treatment of choice. In recurrent disease, topoisomerase I topotecan, the alkylating agent altretamine, and liposomal doxorubicin are used as a monotherapy.

Testicular Cancer

In advanced testicular cancer, platinum-based therapy is the effective treatment. PEB protocol consists of three cycles of cisplatin, etoposide, and bleomycin, or four cycles of cisplatin and etoposide show better results. In patients with high-risk disease, the combination of cisplatin, etoposide, and ifosfamide can be used as well as etoposide and bleomycin with high-dose cisplatin.

Malignant Melanoma

This is a drug-resistant tumor, and it is difficult to treat because it spreads to metastatic sites. Interferon-α and interleukin-2 (IL-2) have greater activity against this melanoma.

Brain Cancer

Nitrosoureas cross the blood–brain barrier and are the most active agents against brain cancer. Carmustine or lomustine are used alone or in combination with procarbazine and vincristine. The anti-VEGF antibody bevacizumab alone or in combination with chemotherapy has promising activity in adult glioblastoma multiforme (Deck and Winston 2012).

FDA-Approved Drugs

Name of Drug	Indications
Osimertinib	Metastatic EGFR T790M mutation—positive NSCLC, as detected by FDA-approved test, progressing during or after EGFR TKI therapy
Daratumumab (Darzalex)	Multiple myeloma after three or more prior lines of therapy, including immunomodulatory agent
Ixazomib	In combination with lenalidomide and dexamethasone for multiple myeloma after one or more prior therapies
Necitumumab	In combination with gemcitabine and cisplatin for first-line treatment of metastatic squamous NSCLC
Alectinib	ALK-positive metastatic NSCLC progressing with or intolerant to crizotinib
Venetoclax	CLL with 17p deletion, as detected by FDA-approved test, after one or more prior therapies
Atezolizumab	Locally advanced or metastatic urothelial carcinoma progressing during or after platinum-containing chemotherapy or within 12 months of neoadjuvant or adjuvant treatment with platinum-containing chemotherapy
Palbociclib	CDK4/6 inhibitors are a new class of targeted treatments that may have a role in breast cancer
Inotuzumab ozogamicin	Antibody–drug conjugates to improve outcomes after acute lymphoblastic leukemia relapse
Alectinib	Provides good results in patients with crizotinib-resistant, advanced NSCLC, including those with brain metastases
Cabozantinib	Oral treatment that blocks several different targets in cancer cells, including tyrosine kinases MET, VEGFR2, and AXL, and used in kidney cancer
Mirvetuximab soravtansine	Ovarian cancer (underway)
Pembrolizumab	PD-L1-positive advanced gastric cancer and treatment of patients with recurrent or metastatic head and neck squamous cell carcinoma (HNSCC) with disease progression on or after platinum-containing chemotherapy
Regorafenib	Treatment of patients with hepatocellular carcinoma (HCC) who have been previously treated with sorafenib
Niraparib	A poly ADP-ribose polymerase (PARP) inhibitor, for the maintenance treatment of adult patients with recurrent epithelial ovarian, fallopian tube, or primary peritoneal cancer who are in complete or partial response to platinum-based chemotherapy
Avelumab	A programmed death-ligand 1 (PD-L1) blocking human IgG1 lambda monoclonal antibody used to treat metastatic Merkel cell carcinoma
Ribociclib	A cyclin-dependent kinase 4/6 inhibitor, in combination with an aromatase inhibitor in a cyclin-dependent kinase 4/6 inhibitor, in combination with an aromatase inhibitor
Lenalidomide	Maintenance therapy for patients with multiple myeloma following autologous stem cell transplant
Cabozantinib	Treatment of advanced renal cell carcinoma
Venetoclax	Treatment of chronic lymphocytic leukemia
Everolimus	Treatment of adult patients with progressive, well-differentiated, nonfunctional, neuroendocrine tumors (NET) of gastrointestinal (GI) or lung origin with unresectable, locally advanced or metastatic disease

(Continued)

Name of Drug	Indications
Eribulin	Treatment of patients with unresectable or metastatic liposarcoma who have received a prior anthracycline-containing regimen
Ofatumumab	Treatment of progressive chronic lymphocytic leukemia
Nivolumab	Treatment of patients with locally advanced or metastatic urothelial carcinoma who have disease progression during or following platinum-containing chemotherapy or have disease progression within 12 months of neoadjuvant or adjuvant treatment with a platinum-containing chemotherapy
	Treatment of classical Hodgkin's lymphoma (cHL) that has relapsed or progressed after autologous hematopoietic stem cell transplantation (HSCT) and posttransplantation brentuximab vedotin (Adcetris)
Obinutuzumab	Treatment of patients with follicular lymphoma (FL) who relapsed after, or are refractory to, a rituximab-containing regimen; previously approved for use in combination with chlorambucil for the treatment of patients with previously untreated chronic lymphocytic leukemia
Rucaparib	Treatment of patients with advanced ovarian cancer who have been treated with two or more chemotherapies

General Toxicity of Cytotoxic Drugs

Side effects of cytotoxic drugs include nausea, vomiting, alopecia, oligozoospermia, impotence, amenorrhoea, abortion, carcinogenicity, hyperuricemia, and opportunistic infections.

Specific Toxicity of Anticancer Drugs

- Cystitis, alopecia—Cyclophosphamide
- Neuropathy—Vincristine
- Cardiac toxicity—Doxorubicin
- Pulmonary fibrosis—Bleomycin and busulfan
- Cisplatin—Nephrotoxicity
- Methotrexate—Megaloblastic anemia and pancytopenia

REFERENCES

American Cancer Society. Treatment. 2016. https://www.cancer.org/treatment/treatments-and-side-effects.html

Angst E, Hiatt JR, Gloor B, Reber HA, Hines OJ. Laparoscopic surgery for cancer: A systematic review and a way forward. *J Am Coll Sur.* 2010;211(3):412–423.

Apolo AB, et al. Avelumab (MSB0010718C; anti-PD-L1) in patients with metastatic urothelial carcinoma from the JAVELIN solid tumor phase 1b trial: Analysis of safety, clinical activity, and PD-L1 expression. Abstract 4514. ASCO Annual Meeting, Chicago, IL, June 3–7, 2016.

Baskar R, Lee KA, Yeo R, Yeoh KW. Cancer and radiation therapy: Current advances and future directions. *Int J Med Sci.* 2012;9(3):193–199.

Begg AC, Stewart FA, Vens C. Strategies to improve radiotherapy with targeted drugs. *Nat Rev Cancer.* 2011;11:239–253.

Brahmer J, et al. Nivolumab versus docetaxel in advanced squamous-cell non-small-cell lung cancer. *N Engl J Med.* 2015;373(2):123–135.

Bristol-Myers Squibb. Nivolumab combined with Ipilimumab versus Sunitinib in previously untreated advanced or metastatic renal cell carcinoma (CheckMate 214). Available at: https://clinicaltrials.gov/show/NCT02231749.

Deck DH, Winston LG. Antimycobacterial drugs. In: Katzung BG, Masters SB, Trevor AJ, ed. *Basic and Clinical Pharmacology.* 12th ed. New York, NY: McGraw-Hill Lange; 2012:839–847.

Doniņa S, et al. Adapted ECHO-7 virus Rigvir immunotherapy (oncolytic virotherapy) prolongs survival in melanoma patients after surgical excision of the tumor in a retrospective study. *Melanoma Res.* 2015;25(5):421–426.

Eggermont AM, et al. Adjuvant ipilimumab versus placebo after complete resection of high-risk stage III melanoma (EORTC 18071): A randomised, double-blind, phase 3 trial. *Lancet Oncol.* 2015;16(5):522–530.

Emens LA. Trastuzumab: Targeted therapy for the management of HER-2/neu-overexpressing metastatic breast cancer. *Am J Ther.* 2005;12(3):243–253.

Hammers HJ, et al. Expanded cohort results from CheckMate 016: A phase I study of nivolumab in combination with ipilimumab in metastatic renal cell carcinoma (mRCC). Abstract 4516. ASCO Annual Meeting, Chicago, IL, May 29–June 2, 2015.

Hauspy J, Jiménez W, Rosen B, Gotlieb WH, Fung-Kee-Fung M, Plante M. Laparoscopic surgery for endometrial cancer: A review. *J Obstet Gynaecol Can.* 2010;32(6):570–579.

Hui R, et al. Long-term OS for patients with advanced NSCLC enrolled in the KEYNOTE-001 study of pembrolizumab (pembro). Abstract 9026. ASCO Annual Meeting, Chicago, IL, June 3–7, 2016.

Hynes NE, Lane HA. ERBB receptors and cancer: The complexity of targeted inhibitors. *Nat Rev Cancer.* 2005;5(7):580.

Jackson SP, Bartek J. The DNA-damage response in human biology and disease. *Nature.* 2009;461:1071–1078.

Kahnamoui K, Cadeddu M, Farrokhyar F, Anvari M. Laparoscopic surgery for colon cancer: A systematic review. *Can J Surg.* 2007;50(1):48–57.

Long G, et al. Primary analysis of MASTERKEY-265 phase 1b study of talimogene laherparepvec (T-VEC) and pembrolizumab (pembro) for unresectable stage IIIB-IV melanoma. Society for Melanoma Research 2015 Congress, San Francisco, CA, November 18–21, 2015.

Malinowsky K, Wolff C, Gündisch S, Berg D, Becker K. Targeted therapies in cancer - challenges and chances offered by newly developed techniques for protein analysis in clinical tissues. *Journal of Cancer.* 2011;2:26–35.

Massard C, et al. Safety and efficacy of durvalumab (MEDI4736), an anti-PD-L1 antibody, in urothelial bladder cancer. Abstract 4502. ASCO Annual Meeting, Chicago, IL, June 3–7, 2016.

Mohammadzadeh N, Safdari R. Robotic surgery in cancer care: Opportunities and challenges. MINI-REVIEW. *APJCP.* 2014;15(3):1081–1083.

National Cancer Institute. Cryosurgery. 2015. https://www.cancer.gov/about-cancer/treatment/types/surgery.

Orihuela E, Smith AD. Laser treatment of transitional cell cancer of the bladder and upper urinary tract. *Cancer Treat Res.* 1989;46:123–142.

Piccart-Gebhart MJ, Procter M, Leyland-Jones B, et al. Trastuzumab after adjuvant chemotherapy in HER2-positive breast cancer. *N Engl J Med.* 2005;353:1659–1672.

Plimack ER, et al. Pembrolizumab (MK- 3475) for advanced urothelial cancer: Updated results and biomarker analysis from KEYNOTE-012. Abstract 4502. ASCO Annual Meeting, Chicago, IL, May 29–June 2, 2015.

Puzanov I, et al. Survival, safety, and response patterns in a phase 1b multicenter trial of talimogene laherparepvec (T-VEC) and ipilimumab (ipi) in previously untreated, unresected stage IIIB-IV melanoma. Abstract 9063. 2015 ASCO Annual Meeting, Chicago, IL, May 29–June 2, 2015.

Redman JM, et al. Advances in immunotherapy for melanoma. *BMC Med.* 2016;14(1):20.

Schadendorf D, et al. Pooled analysis of long-term survival data from phase II and phase III trials of ipilimumab in unresectable or metastatic melanoma. *J Clin Oncol.* 2015;33(17):1889–1894.

Sheridan AT, Dawber RP. Curettage, electrosurgery and skin cancer. *Australas J Dermatol.* 2000;41(1):19–30.

Smith DA, et al. Updated survival and biomarker analyses of a randomized phase II study of atezolizumab vs docetaxel in 2L/3L NSCLC (POPLAR). Abstract 9028. ASCO Annual Meeting, Chicago, IL, June 3–7, 2016.

Subotic S, Wyler S, Bachmann A. Surgical treatment of localised renal cancer. *Eur Urol Suppl.* 2012;11(3):60–65.

Tatli S, Tapan U, Morrison PR, Silverman SG. Radiofrequency ablation: Technique and clinical applications. *Diagn Interv Radiol.* 2012;18:508–516.

Whitson BA, Groth SS, Duval SJ, Swanson SJ, Maddaus MA. Surgery for early-stage non-small cell lung cancer: A systematic review of the video-assisted thoracoscopic surgery versus thoracotomy approaches to lobectomy. *Ann Thorac Surg.* 2008;86(6):2008–2016.

5 Bacteria and Bioactive Peptides
A Robust Foundation to Combat Cancer

Ameer Khusro, Chirom Aarti, and Paul Agastian

CONTENTS

INTRODUCTION

Cancer is one of the devastating noncommunicable diseases that remain leading causes of mortality as well as morbidity in this industrialized world. In general, cancer has been defined as uncontrolled and invasive growth of cells that leads to the deadliest side effects. Cancer or tumor cells are well known for their unique characteristics (i.e., escaping normal growth-regulating processes). In fact, there is a balance between cell formation and cell death under normal growth conditions, causing a constant number of a particular cell type. But during tumor formation or neoplasm, cells do not respond to the normal growth control mechanisms and give rise to uncontrolled growth of cells. Six essential alterations in cell physiology are shown by tumor cells: self-sufficiency in growth signals, growth-inhibitory signal insensitivity, tolerance to programmed cell death, limitless replicative ability, sustained angiogenesis, and metastasis (Hanahan and Weinberg 2011). In 2012, globally 14.1 million new cases of cancer were estimated, and 8.2 million mortalities due to cancer were reported (Ferlay et al. 2013). In fact, at present, cancer biology is one of the most investigated ongoing research avenues because of its uncontrolled spreading and millions of mortalities due to the lack of suitable treatment.

Conventional anticancer therapies include surgery, radiotherapy, and chemotherapy, which are not very effective in the treatment of solid tumors, and these therapies are often associated with serious side effects. Today, we are in search of alternative techniques to target tumors in a nontoxic or less toxic manner.

The exploitation of bacteria for the treatment of tumors is not new. In 1868, W. Busch tried to cure a female patient by first cauterizing her neck tumor and subsequently placing the woman into a bed, which had previously been occupied by a patient with "erysipelas," a *Streptococcus pyogenes* infection (Pawelek et al. 2003). This resulted in rapid reduction in tumor growth. After this occurrence, several clinicians exploited various bacteria and their respective constituents for the treatment of tumors. A mixture of inactivated *Streptococci* and *Serratia marcescens*, called "Coley's toxin," is still well known to oncologists (Coley 1991).

The live, attenuated, or genetically modified, nonpathogenic bacteria or their tumoricidal constituents emerged as potent antitumor agents for the treatment of this disease. Later studies used pathogenic species of the anaerobic clostridia for their proliferation within the necrotic (anaerobic) regions of tumors in experimental animals as compared to normal tissues. The experiments caused regression of tumors but showed acute toxicity as well as death of most of the experimented animals (Minton 2003). These drastic outcomes shifted to the use of a nonpathogenic strain of *Clostridium*, which showed its potentiality to colonize anaerobic parts of the tumor after intravenous administration, but the outcome was not satisfactory in terms of tumor regression (Carey et al. 1967). Anaerobic bacterial species such as *Clostridium novyi* demonstrated significant antitumor activity but showed lethal effects on experimental animals. The attenuated *C. novyi*-NT exhibited significant antitumor effects but also showed severe toxicity. The use of obligate anaerobic bacteria or facultative anaerobic as alternative cancer therapeutics has been followed over more than a century due to the insights gained over time on host–bacteria interactions, genome information for many of the bacteria employed, and especially efficient molecular tools. A number of bacterial species have proven their tumor therapeutic effects in murine model systems. The lack of effectiveness of these bacterial therapies encouraged continuous efforts to identify other novel bacterial strains as anticancer therapeutic tools.

Pathogens frequently cause cell death in infected tissues. Apoptosis, or programmed cell death, is the principal physiological process for eliminating unnecessary cells during embryogenesis, in the development and regulation of the immune system, and during normal cell turnover in many tissues (Marsden et al. 2002). Necrosis is usually associated with tissue injury from an interruption in the vascular supply, decreased oxygenation, or various noxious environmental factors (Majno and Joris 1995). *Bifidobacterium breve* and *Bifidobacterium adolescentis* have also been shown to target model tumors and act as reliable gene therapy vectors (Hidaka et al. 2007).

A number of facultative bacteria such as *Escherichia coli*, *Shigella flexneri*, *Vibrio cholera*, and *Listeria monocytogenes* have been employed for their potential role in bacteria-mediated tumor therapy. The facultative anaerobes can colonize hypoxic and necrotic areas as well as viable cells in tumors because they are not restricted to hypoxic environments (Leschner and Weiss 2010). *S. flexneri*, *V. cholera*, and *L. monocytogenes* showed colonization in various model tumors in mice but exhibited a lack of tumoricidal activity (Yu et al. 2008). The findings, therefore, suggest that several bacteria are able to colonize tumors, but possessing antitumor activity is still limited. *E. coli* DH5α has been shown to grow in various subcutaneous tumors of nude mice, and it is able to grow in small metastatic nodules. Likewise, *E. coli* strain TOP10, strain CFT073, the enteroinvasive strain 4608–58, and Nissle 1917 have been shown to colonize solid tumors (Stritzker et al. 2007).

BACTERIA AS CARRIERS OF ANTITUMOR AGENTS

The treatment of liver cancer in experimental models was carried out by engineering a cya/crp (genes encoding proteins involved in the regulation of cyclic AMP levels) mutant of *S. typhimurium* ×4550, and expressing the interleukin-2 (Saltzman et al. 1996, 1997). In view of the colonizing property of *S. typhimurium* in the liver, it is hypothesized that its attenuated form could be employed to deliver cytokines locally to the liver, with an effect on hepatic metastases. Previously, TNF-α and platelet factor 4 fragment had been cloned and expressed in VNP20009 (Lin et al. 1999; Karsten et al. 2001). But, hIL-12, hGM-CSF, mIL-12, and mGM-CSF had been cloned under the control of a cytomegalovirus (CMV) promoter into SL3261, an auxotrophic *S. typhimurium*. TNF-α has been cloned and expressed in *C. acetobutylicum*. *B. adolescentis* was used as a delivery system for the anti-angiogenic protein endostatin and reduced tumor growth (Li et al. 2003). Bacteria can also be used as vectors for preferentially delivering antitumor agents, cytotoxic peptides, therapeutic proteins, or prodrug converting enzymes to solid tumors, which provides an adjuvant therapy.

BACTERIA-MEDIATED ENZYME PRODRUG THERAPY

This therapy uses anaerobic bacteria that have been transformed with a specific enzyme that converts a nontoxic prodrug into a toxic drug. The enzyme is expressed solely due to the proliferation of the bacteria in the necrotic and hypoxic regions of the tumor and thus, the prodrug is metabolized to the toxic drug only in the tumor. Cytosine deaminase (CD) converts 5-fluorocytosine (5FC) to 5-fluorouracil (5FU), and nitroreductase (NR) converts the prodrug CB1954 to a DNA crosslinking agent. These enzymes have been tested with *Clostridium sporogenes*, showing satisfactory *in vitro* results, but *in vivo* results are disappointing. Recently it was illustrated that CD can be successfully cloned and expressed in the same strain of *Clostridium*. CD expression was enhanced significantly by combretastatin A-4 phosphate that might be due to the enlargement of the necrotic area in tumors

(Theys et al. 2001). Other *in vivo* studies include the incorporation of NR and CD with *Salmonella* vector, which is undergoing phase I clinical trials in cancer patients. TAPET (Tumour Amplified Protein Expression Therapy) uses an attenuated strain of *S. typhimurium* (VNP20009) as a bacterial vector and expresses an *E. coli* CD in order to deliver anticancer drugs to solid tumors (Luo et al. 2001). The expression of the prodrug-converting enzyme HSV-thymidine kinase (TK) in a purine auxotroph showed enhanced anticancer activity upon the addition of the corresponding prodrug (Pawelek et al. 1997). Transfected *B. longum* by pBLES100-S-eCD produces cytosine deaminase in the hypoxic tumor, suggesting an effective prodrug-enzyme therapy (Fujimori et al. 2003).

BACTERIAL TOXINS AND SPORES FOR ANTITUMOR THERAPY

Bacterial toxins have also been known to treat cancer by altering the cellular processes that maintain proliferation, apoptosis, and differentiation. Cytolethal distending toxins (CDTs) and the cycle-inhibiting factor (Cif) are cell-cycle inhibitors that block mitosis and compromise the immune system by inhibiting clonal expansion of lymphocytes. In contrast, the cytotoxic necrotizing factor (CNF), a cell-cycle stimulator, promotes cellular proliferation and interferes with cell differentiation (Nougayrede et al. 2005). CNF is a cell-cycle stimulator released by *E. coli*. CNF induces G1-S transition and causes DNA replication. CDTs are present in *Campylobacter jejuni* and *S. typhi*, while Cif is found in enteropathogenic (EPEC) and enterohemorrhagic (EHEC) *E. coli*. The anticancer activity of toxins shows reduced side effects compared to traditional treatments. Further the antitumor activity can be enhanced by combining bacterial toxins with anticancer drugs or irradiation (Carswell et al. 1975).

Diphtheria toxin (DT) is known to bind with the surface of cells, expressing the heparin-binding epidermal growth factor (HB-EGF) precursor, forming a DT-HB-EGF complex. DT undergoes post-translational modifications, forming DT fragment A. This catalytically ribosylates elongation factor-2 (EF-2) leading to inhibition of protein synthesis with subsequent cell lysis and/or induction of apoptosis (Lanzerin et al. 1996; Louie et al. 1997; Falnes et al. 2000; Frankel et al. 2002). Likewise, *Pseudomonas* exotoxin A is also known to inhibit protein synthesis.

Clostridium perfringens type A strain produces *Clostridium perfringens* enterotoxin (CPE) whose C-terminal domain shows high-affinity binding to the CPE receptor (CPE-R), and the N-terminal is assumed to be essential for cytotoxicity (Kokai Kun and Mcclane 1997; Kokai Kun et al. 1999). The purified CPE exerts an acute cytotoxic effect on pancreatic cancer cells and led to tumor necrosis as well as tumor growth inhibition *in vivo*. It is being investigated for colon, breast, and gastric cancers. Recently, botulinum neurotoxin (BoNT) showed more effective destruction of tumor cells (Ansiaux and Gallez 2007). At present, Alfa-toxin from *Staphylococcus aureus*, AC-toxin from *Bordetella pertussis*, shiga-like toxins, and cholera toxin are being investigated on mesothelioma cells (P31) and small lung cancer cells (U-1690). The study showed that AC-toxin caused increased cytotoxicity and apoptosis in a dose-dependent manner against both cell lines. However, the findings of cholera toxins are not satisfactory.

Pseudomonas exotoxin, diphtheria toxin, and ricin may be potent anticancer agents, but these toxins need to be targeted to specific sites on the surface of tumor cells. This process is achieved by conjugating the toxins to cell-binding proteins such as monoclonal antibodies, and thus killing the cancer cells without affecting the normal cells, which do not bind the conjugates. DT ligands such as IL-3, IL-4, granulocyte colony stimulating factor (G-CSF), transferrin (Tf), EGF, and vascular endothelial growth factor (VEGF) have been investigated for targeted cancer cells (Frankel et al. 2002). Few of them have reached different phases of clinical trial.

Recombinant toxins are another effective approach for cancer treatment that includes deleting the DNA coding for the toxin binding region and replacing it with various complementary DNA encoding other cell binding proteins. This approach may be useful in the future for designing toxin-based anticancer treatments. For targeted DT therapy, deletions within the DT receptor binding domain (amino acid residues 390–535) or targeted mutations of the critical HB-EGF precursor binding loop (amino acid residues 510–530) have been used (Greenfield et al. 1987; Frankel et al. 2002). A recombinant interleukin-4- *Pseudomonas* exotoxin (IL4-PE) has been developed for the treatment of glioblastoma, and also showed significant anti-cancer properties against the human glioblastoma tumor model. In addition to this, IL4-PE is being investigated for the treatment of malignant astrocytoma (Puri 1999).

Spores of genetically modified strain, *C. novyi-NT*, devoid of the lethal toxin have shown targeted anticancer action without any systemic side effects. An intratumoral and intravenous injection of *C. histolyticum* and *C. sporogenes* spores, respectively, caused marked lysis of cancer cells. Additionally, *Clostridium* was detected only in cancer cells and not in normal tissues of mice after intravenous injection (Thiele et al. 1963). Further, the study demonstrated that *C. novyi-NT* spores were rapidly cleared from the circulation by the reticuloendothelial system without showing any clinical toxicity. Bacterial spores have also been used as delivery agents for anti-cancer agents, cytotoxic peptides, and therapeutic proteins, and as vectors for gene therapy.

SALMONELLA: A POTENT ANTICANCER AGENT

In spite of significant anticancer activity of several bacterial strains, *Salmonella* sp. is known to play considerable role as an anticancer agent, including direct attachment, invasion, and destruction of tumor cells. Bacteria act as antigens for the production of antibody to specific tumor components. *Salmonella enterica* strains are facultative intracellular enteric pathogens that can produce localized enteritis and disseminated systemic disease in humans and a variety of other vertebrates (Ohl and Miller 2001). The bacterium is also pathogenic to a variety of domestic and wild animals. The bacteria initially infect the intestinal tract and cause widespread destruction of the intestinal mucosa. In addition to this, *Salmonella* strains can also produce serious and fatal infections with considerable cytopathology in a number of systemic organs using specific virulence mechanisms to induce host cell death. *Salmonella* rapidly disseminate to systemic organs of the reticuloendothelial system by a mechanism dependent on CD18-positive cells (Vazquez-Torres et al. 1999). The bacteria attach to the intestinal epithelial cells and induce uptake of the bacteria into specialized membrane-bound vesicles called *Salmonella*-containing vacuoles (SCVs) (Knodler and Steele-Mortimer 2003), which requires the function of a type III protein secretion system (TTSS) encoded in the *Salmonella* pathogenicity island-1 (SPI-1) locus (Galan 2001).

In vivo studies have shown the tumor-controlling property of *Salmonella* in the gut. It has also been observed that cancer patients sometimes get better after they have been exposed to an infection. In this regard, *S. typhimurium* has been extensively studied in order to target tumors. Most importantly, the bacteria were shown not only to colonize large established cancer cells but also to exhibit the property of invasion and affecting metastases (Guiney 2005). These bacteria are aerobic and anaerobic that can invade epithelial cells, macrophages, and dendritic cells of the host. Most importantly, *Salmonella* sp. have the unique property to preferentially colonize solid tumors, which cause retardation in tumor growth, rendering (Pawelek et al. 1997) *Salmonella* an effective candidate for bacteria-mediated tumor therapy.

The side effects, specificity, and a systemic delivery procedure to reach solid tumors are the ideal factors to be considered in anticancer strategy. The major advantages of developing *Salmonella* strains as a delivery system are preferred migration to tumor cells versus nontumor cells, nontoxicity, control with antibiotics, a large genomic reservoir for strain selection, engineering stimulation of the immune system at the tumor site, and the ability to deliver "foreign" genes by plasmid transfection and phage transduction. Properties viz. motility, facultative anaerobiosis, and the potentiality to invade epithelial cells and engineered auxotrophies may contribute to successful interference of tumor (Moreno et al. 2010). This bacterium showed long-term anticancer effect against several experimental tumors and showed the ability to target metastatic lesions (Low et al. 1999; Luo et al. 1999).

S. typhimurium strain VP20009 showed nontoxicity in humans, and was administered systemically to colon cancer and melanoma patients in phase I trials with fewer side effects (Nemunaitis et al. 2003). In fact, this strain is attenuated, which reduces the toxic effects of its lipopolysaccharide and creates a requirement for an external source of adenine. A tumor-targeting strategy for pancreatic cancer using a modified auxotrophic strain of *S. typhimurium* was also reported (Nagakura et al. 2009). The genetically modified strain of *S. typhimurium* requires the amino acids arginine and leucine. *S. typhimurium* A1-R and expressing green fluorescent protein (GFP) were administered to an orthotopic human pancreatic tumor expressing red fluorescent protein (RFP) in nude mice. After a required period of treatment, the regression in the pancreatic cancer was observed without any further chemotherapy treatment, suggesting the potential of bacteria targeting pancreatic cancer.

Salmonella attack cancer cells in multiple ways. The bacteria preferentially accumulate at the region of the tumor and could stimulate local nonspecific immune responses (Agrawal et al. 2004). *Salmonella* could compete with cancer cells for nutrients (Zhao et al. 2005), thus impeding further growth of tumor cells. The bacterial lysis attached to tumors might release antitumor cellular components. Engineered *Salmonella* introduces genetic vectors that express antitumor constituents. Salmonella-induced necrosis releases tumor peptides that could also be used to stimulate antitumor antibodies production. Rapid growth of *Salmonella* in necrotic and hypoxic regions may allow *Salmonella* selective multiplication in tumors. *Salmonella* might be used to specifically express peptides at tumor regions, demonstrating innate antitumor functions. An auxotrophic mutant of *Salmonella* had been used in a mouse xenograft model of human prostate cancer with multiple bacterial doses (Zhao et al. 2007). The successful future use of *S. typhimurium* in cancer therapy may be aided by finding the optimal synergy between these direct killing mechanisms and the engineered capacity to target molecular pathways that are critical to tumor survival.

Salmonella sp. can induce apoptosis or necrosis by direct and indirect mechanisms including both exogenous and endogenous apoptosis, tumor necrosis factor, and FAS receptors (Weinrauch and Zychlinsky 1999). The infection of the macrophages cell line with attenuated *S. typhimurium* caused a significant enhancement in lipid peroxidation and lowered the antioxidants activities, which correlated with the increased generation of tumor necrosis factor-α, interleukine-1α, interleukine-6, and triggered caspase-3 apoptosis, suggesting that caspase-3 and tumor necrosis factor-α in conjunction with other cytokines may induce apoptotic cell death in *Salmonella* infected cells through the upregulation of oxidants and downregulation of antioxidants (Chanana et al. 2007). In another report, it was indicated that increased levels of nitrite and decreased levels of superoxide dismutase may be one of the factors to induce apoptosis in *Salmonella* infected cells (Chander et al. 2006). *S. typhimurium* induces apoptosis in infected macrophages within 14 hours from the time of infection through caspase-1-dependent mechanism (Takaya et al. 2005). Genes in the chromosomally encoded *Salmonella* pathogenicity island 2 (SPI 2) that encodes a Type III secretion system and genes on the virulence plasmid pSDL2 of *Salmonella enteritica* serovar Dublin (*spv* genes) were found to have a significant role in determining the ultimate fate of infected cells with respect to caspase 3 activation and undergoing death by apoptosis (Paesold et al. 2002). Vendrell et al. (2010) examined the mechanisms mediating the Th-1 inducing effect of *S. typhi* at solid tumors. They found that *Salmonella* promoted an antitumor Th1-type response characterized by increased frequencies of IFN-γ-secreting CD4 (+) T and CD8 (+) T cells with reduction of regulatory T cells in tumor-draining lymph nodes. They also found that the main cells infiltrating bacteria-treated tumors were activated neutrophils, which can exert an antitumor effect through the secretion of TNF-α.

The use of attenuated *Salmonella* strains as live vectors showed broad applications in medicine, especially in cancer treatment. The majority of solid tumors contain regions of low oxygen or dead tissue, which encourages the growth of *Salmonella* making them ideal vectors to deliver anticancer components. The property such as prolonged colonization of tumors makes *Salmonella* sp. a better treatment over recombinant viral cancer gene therapy. However, the safety and efficiency of gene transfer and duration of expression are limited for this bacterial therapy approach (Basu and Herlyn 2008). *S. typhimurium* has been engineered in order to express cytosine deaminase (Nemunaitis et al. 2003), which converts administered nontoxic 5-fluorocytosine to the active chemotherapeutic agent 5-fluorouracil. *S. typhimurium* has also been assessed as a potent immunotherapy vector through expression of human tumor antigens (Fensterle et al. 2008). *S. typhimurium* is used as both a vehicle for DNA tumor vaccines (Bauer et al. 2005) and an antitumor gene therapy vector (Yoon et al. 2011) due to its ability to transfer eukaryotic expression plasmids to mammalian cells. Additionally, *Salmonella* sp. have been used as delivery systems for tumor-associated antigens in tumor immunotherapy.

Human telomerase reverse transcriptase promoter was inserted in *S. typhimurium* to confine gene expression strictly to the telomerase-positive tumors (Fu et al. 2008). Attenuated *S. typhimurium* was used as a carrier to express mitochondria-derived activator of apoptosis caspases, or smac, and TRAIL genes under the control of the human telomerase promoter. The *in vitro* reports showed that Smac could enhance TRAIL-induced apoptosis in cancer cells, and the human telomerase promoter could drive specific gene expression in cancer cells but not in normal cells. But *in vivo* results showed that salmonella-mediated exogenous gene expression could persist for at least 14 days in tumors and significantly inhibited the growth of tumor in mice.

BACTERIAL PROTEINS: AZURIN AS POTENTIAL ANTICANCER DRUGS

Azurin can enter cancer cells much more preferentially than normal cells and interferes in tumor growth by multiple modes of action including complex formation with the tumor suppressor protein p53 (Bizzarri et al., 2011), stabilizing it and enhancing its intracellular level, which then allows induction of apoptosis uniquely in tumors where it entered, leading to cancer cell death and shrinkage in mice (Yamada et al. 2004). In fact, azurin inhibits angiogenesis in tumors through inhibition of the phosphorylation of VEGFR-2, FAK, and AKT22. Azurin also has other tumor growth inhibitory activities that p28 lacks, such as azurin does not enter the cancer cells to form complexes with p53, VEGFR, FAK, and AKT in order to inhibit their functions. Azurin has the property to target cancer cells growing by expressing certain cell signaling receptor tyrosine kinase molecules on the cell surface. For example, a receptor kinase EphB2 is hyperproduced at the surface of cancer cells such as breast, prostate, lung, etc., promoting their rapid growth and proliferation when bound with its cell-membrane-associated ligand ephrin B2. Most importantly, azurin preferentially enters cancer cells and forms complexes with key proteins responsible for tumor formation, finally inhibiting the growth of cancer cells. Azurin has C-terminal four loop regions termed *CD loop, EF loop, FG loop,* and *GH loop* as well as its structural similarity with antibody variable domains of various immunoglobulins giving rise to a β-sandwich core and an immunoglobulin fold. This unique structure allows azurin to evade immune action and depicts its anticancer property when present in the bloodstream (Fialho et al. 2007). Azurin's binding domain to EphB2 via its G-H loop region has been used to increase radiation sensitivity of lung tumor cells through conjugation with the radio-sensitizer nicotinamide (Micewicz et al. 2011). Azurin has the ability to inhibit the growth of highly invasive P-cadherin (a member of the type I cadherin family that in certain conditions acts not as a regular cell–cell adhesion molecule, but as a promoter for malignant breast tumor progression) overexpressing breast cancer cells (Bernardes et al. 2013). A sublethal single dose of azurin produced a decrease in the invasion of two P-cadherin expressing breast cancer cell models, the luminal MCF-7/AZ.Pcad and the triple negative basal-like SUM 149 PT through a Matrigel artificial matrix.

The decrease in invasion corresponds to a decrease in the total P-cadherin protein levels and a concomitant decrease of its membrane staining, whereas E-cadherin remains unaltered with high expression levels and with normal membrane localization, indicating the very important role of azurin only for invasive cell lines (Bernardes et al. 2013). Despite the potentiality of azurin to inhibit these proteins, the exact mechanism remains elusive.

In fact, Azurin interferes with signaling pathways associated to cancer Src and FAK are nonreceptor tyrosine kinases that mediate several biological processes associated with cell adhesion, migration, and invasion (Luo and Guan 2010). Several cancer models show deregulation in signaling from these molecules. Src and FAK were identified as mediators of the crosstalk between integrin- and cadherin-mediated adhesion in epithelial cells and other cell surface receptor proteins (Goel et al. 2012). The alterations mechanisms of azurin in biological processes associated with a more aggressive phenotype in invasive cancer cells were further explored in a transcriptomic profiling of invasive P-cadherin-overexpressing MCF-7/AZ (Bernardes et al. 2014). Azurin upregulates genes associated to apoptosis. It also downregulates genes responsible for endocytosis via vesicle-mediated transport and with lysosomal degradation. In general, azurin acts by decreasing membrane receptors and their hyperactivated signaling pathways that may be sustaining tumor growth.

BACTERIAL PEPTIDES AS POTENT ANTICANCER AGENTS

Peptides comprise a few amino acids to about 40 or more amino acids coupled through amide and/or disulfide bonds, providing varied-size molecules. In the last few years, Leuprorelin, Octreotide, and Goserelin have been used against prostate cancer and breast cancer. Carfilzomib, a protease inhibitor, is used for multiple myeloma (Kaspar and Reichert 2013). The administration of peptides can be oral, subcutaneous, intravenous, or via inhalation. In spite of limited anticancer peptides, peptide therapy may have significant potential in tumor treatment. A synthetic six amino acid peptide of αC-ß4 loop region of EGFR has been known to inhibit the dimerization and signaling activity of EGFR in the presence of its ligand (Ahsan et al. 2013). Additionally, this peptide promotes EGFR interaction with the heat shock protein Hsp90, thereby catalyzing the degradation of EGFR (Ahsan et al. 2013). At present, anticancer peptides such as Cilengitide, Trebananib, NGR-hTNF, Tyroserleutide, etc., are undergoing phase III clinical trials against glioblastoma, ovarian, mesothelioma, or liver cancers (Kaspar and Reichert 2013). The lack of targeting intracellular proteins is the major limitation of currently available anticancer peptides, thereby limiting their effectiveness (Milletti 2012). Thus, an ideal anticancer peptide involves the development of cell-penetrating peptides (CPPs) that can cross the cellular membrane to modulate key intracellular proteins involved in cancer growth regulation.

A helical peptide with a stretch of hydrophobic amino acids, termed p18, and an extended form of p18 with 10 additional amino acids, termed p28 have the significant property of being able to preferentially invade tumors (Taylor et al. 2009). P28 is a component of azurin, which enters preferentially to cancer cells. After invasion, p18 has the protein transduction domain (PTD) with a lower anticancer property. Thus, tumors are located inside the body using fluorescently labeled p18 as a diagnostic marker where it accumulates due to selective entry.

P28 preferentially enters the cancer cells and forms a complex within the p53 DNA binding domain, causing inhibition of ubiquitination and proteasomal degradation via an HDM-2 independent pathway. Apoptosis and cell cycle arrest in breast and other cancer cells occurs due to the inhibition of p53 degradation (Yamada et al. 2009). Apart from this, p28 inhibits angiogenesis in cancer cells by inhibiting phosphorylation of VEGFR-2, FAK, and AKT 22, and thus affecting the growth of cancer cells by inhibiting multiple independent pathways.

ROLE OF BACTERIOCINS IN CANCER THERAPY

Bacteriocins are ribosomally synthesized cationic peptides that are produced by almost all groups of bacteria and exhibit biological activities at lower concentrations. Bacteriocins are known to exhibit anticancer activity against several cell lines (Table 5.1).

The cell membrane of cancer carries a predominantly negative charge due to the presence of the anionic phosphatidylserine, O-glycosylated mucins, sialylated

TABLE 5.1
List of Potent Bacteriocins Against Various Cancer Cell Lines

Bacteriocin	Source	Class	Cell Lines	References
Colicin E3	*E. coli*	III	P388, HeLa, HS913T	Fuska et al. (1978); Smarda et al. (1978)
Colicin A	*E. coli*	III	HS913T, SKUT1, BT474, ZR75, SKBR3, MRC5	Chumchalova and Smarda (2003)
Microcin E492	*K. pneumoniae*	IIa	Hela, Jurkat, RJ2.25	Hetz et al. (2002)
Pediocin PA-1	*P. acidilactici* PAC1.0	IIa	A-549, DLD-1	Beaulieu (2004)
Nisin	*L. lactis*	I	MCF7, HepG2	Paiva et al. (2011)
Plantaricin A	*L. plantarum* C11	II	Jurkat, GH4, Reh, Jurkat, PC12, N2A, GH4	Zhao et al. (2006); Sand et al. (2007); Sand et al. (2010); Sand et al. (2013)
Smegmatocin	*M. smegmatis* 14468	III	HeLa AS-II, HGC-27mKS-ATU-7	Saito et al. (1979); Saito and Watanabe (1979, 1981)
Pediocin CP2	*P. acidilactici*	IIa	HeLa, MCF7, Sp2/0-Ag14, HepG2	Kumar et al. (2012)
Colicin E1	*E. coli*	III	MCF7, HS913T	Chumchalova and Smarda (2003)
Pediocin K2a2-3	*P. acidilactici* K2a2-3	IIa	HT2a, HeLa	Villarante et al. (2011)
Bovicin HC5	*S. bovis* HC5	I	MCF7, HepG2	Paiva et al. (2011)
Pyocin S2	*P. aeruginosa* 42A	III	HepG2, Im9HeLa, AS-II, mKS-ATU-7	Abdi-Ali et al. (2004); Watanabe and Saito (1980)

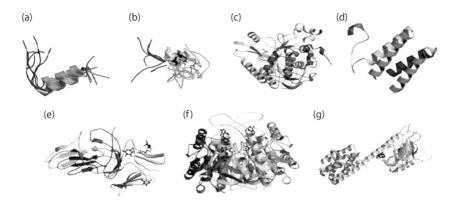

FIGURE 5.1 PDB structure of (a) Plantaricin A, (b) bovicin, (c) nisin, (d) pediocin, (e) pyocins, (f) microcins, and (g) colicins.

gangliosides, and heparin sulfates (Riedl et al. 2011). Bacteriocins bind to negatively charged cell membranes of cancer cells due to their cationic nature (Hoskin and Ramamoorthy 2008). The binding of bacteriocins to cancer cells can be explained due to differences in the membrane fluidity of cancer cells. The membrane of a cancer cell can be easily destabilized due to its higher membrane fluidity as compared to a normal cell (Sok et al. 1999). In addition, the membrane of a cancer cell contains a higher number of microvilli compared to a normal cell, which increases the surface area of a cancer cell, resulting in the binding of several antimicrobial peptides to the cancer cell membrane (Chan et al. 1998). Figure 5.1 shows the PDB structure of biologically active various bacteriocins.

Plantaricin A

Lactobacillus plantarum C11 are known to produce a 2.4 kDa class II type bacteriocin, called Plantaricin A (plnA). It is available in three different forms viz. a 26 residue peptide and two N-terminally truncated forms containing 23 and 22 residues. These three variants are from a 48-residue precursor encoded by plnA gene (Diep et al. 1996). Zhao et al. (2006) demonstrated the factor-dependent cytotoxicity of artificially synthesized plnA against human T-cell leukemia cell line, which showed significant reduction in cell viability at a definite temperature. The microscopic observation of cell nuclei fragmentation and membrane of tumors suggested apoptosis and necrosis as a mode of cell killing. The intracellular concentration of caspase-3 was also found to be increased after plnA treatment. Further, plnA affected the permeability of the liposomes containing negatively charged phospholipids, resulting in leakage of fluorescent dye carboxy fluoresce. Additionally, plnA forms amyloid-like fibrils with phosphatidyl serine (Zhao et al. 2006). The effectiveness of these peptides at lower concentrations is due to the concentrating effect of phospholipid. Previously a significant exposure of phosphatidyl serine on the surface of cancer cells has been studied that showed significant interaction of plnA with cancer cells (Riedl et al. 2011). Sand et al. (2007) reported the effect of artificially synthesized plnA on the permeabilization of normal as well as cancerous rat anterior pituitary cells using whole-cell patch-clamp recordings of the membrane potential and also by microfluorimetry that measured cytosolic Ca^{2+} concentration. However, in another report, Sand et al. (2010) did not demonstrate any significant effect of plnA on normal and cancer cell lines. The cytotoxicity of plnA was reported on normal human B and T cells, Reh cells (human B cell leukemia cell line), Jurkat cells, normal rat cortical neurons and glial cells, PC12 (rat adrenal chromaffin tumor), and murine N2A cell lines (spinal cord tumor cell line). A nonsignificant death of normal and cancer cells was observed in a dose-dependent manner of plnA under physiological conditions. Thus, the structural modifications of plnA need to be tested for their selectivity to various tumors. Sand et al. (2013) reported that in the case of eukaryotic cell membrane, the membrane permeabilizing property of plnA depends on the negative surface charge conferred by glycosylated proteins of membrane. The removal of carbohydrate residues from glycosylated membrane proteins by exposure to PNGase F, chondroitinase ABC, and heparinise I, II, III made the cells resistant to plnA-induced permeabilization, suggesting a significant role of glycosylated membrane proteins in plnA-induced membrane permeabilization.

BOVICIN AND SMEGMATOCIN

Bovicin is a lantibiotic produced by *Streptococcus bovis* HC5. The peptide has a molecular weight of 2.4 kDa and shows stability at low pH as well as very high temperature. The structure and function of this peptide are quite similar to nisin with broad-spectrum biological activities. The biological activity of bovicin is resistant to proteinase K and chymotrypsin but susceptible to pronase E and trypsin. *In vitro* cytotoxicity of bovicin HC5 was observed against MCF7and HepG2 cell lines with IC_{50} values 279.39 and 289.3 mM, respectively (Paiva et al. 2012).

Smegmatocin 14468 is a 75 kDa bacteriocin produced by *Mycobacterium smegmatis* 14468 with narrow-spectrum biological activities. The peptide is heat labile and loses its property at high temperature as well as in the presence of pepsin, trypsin, and chymotrypsin (Saito et al. 1979). The study also reported the inhibition of human cell line HeLaS3, showing morphological changes in the cancer cells such as shrinking and appearance of vacuoles in their cytoplasm (Saito et al. 1979). In another report, treatment of mKS-ATU-7 cells with lower concentrations of smegmatocin showed significant reduction in the total counts of cells as compared to normal untransformed cells due to the reduction in protein and DNA synthesis (Saito and Watanabe 1981). According to Saito and Watanabe (1979), the smegmatocin showed dose-dependent cytotoxicity on AS-II and HGC-27 cell lines. Further studies are required to determine the mode of action of smegmatocin against these human cancer cell lines *in vitro* as well as *in vivo*, because smegmatocin is quite a large protein.

NISIN

Nisin is a low molecular weight heat-stable pentacyclic peptide containing 34 amino acid residue, which is produced by *Lactococcus lactis* (Hansen and Liu 1990).

It contains uncommon amino acids such as lanthinone, methyllanthionine, didehydroalanine, and didehydro-aminobutyric acid, introduced during post-translational modifications of protein (Klaenhammer et al. 1993). Paiva et al. (2012) reported the cytotoxic effect of nisin against MCF-7 and HepG2 cell lines with IC_{50} values of 46 and 112.25 mM, respectively. Maher and McClean (2006) demonstrated the cytotoxicity of nisin against HT29 and Caco-2 cell lines with IC_{50} values of 89.9 and 115 mM, respectively. Joo et al. (2012) reported cytotoxicity of nisin against squamous cell carcinoma (HNSCC) both *in vitro* and *in vivo*. Nisin caused apoptosis by alteration in calcium influx and cell cycle arrest in the G2 phase. Recently Kamarajan et al. (2015) reported the natural variants of nisin, nisin A, and nisin Z for their cytotoxic effects on HNSCC cells using a mice model in a dose-dependent manner. Preet et al. (2015) reported the synergistic impact of nisin along with doxorubicin on dimethylbenz (a) anthracene-induced skin carcinogenesis in mice and showed that nisin and doxorubicin alone or in combination reduced the volume of tumor significantly after a required period of treatment. Cells showed chromatin condensation and marginalization of nuclear material in the nisin-doxorubicin treated group, concluding the ability of nisin to increase the anticancer potential of the chemotherapeutic drug, doxorubicin.

PEDIOCIN

Pediocins are plasmid-encoded thermo-alkali stable class IIa bacteriocins, produced by *Pediococcus* sp. (Papagianni 2003). The peptide is susceptible to several proteolytic enzymes such as papain, pepsin, protease, trypsin, and chymotrypsin (Kumar et al. 2011). Pediocin PA-1 produced by *P. acidilactici* PAC1.0 and the recombinant pediocin cloned in *Pichia pastoris* are known to inhibit the growth of cell lines such as A-549 and DLD-1. Native pediocin PA-1 showed cytotoxicity at lower ranges, whereas the recombinant pediocin was ineffective at lower ranges (Beaulieu 2004). Pediocin isolated from *P. acidilactici* K2a2-3 showed cytotoxicity against HT29, a human colon adenocarcinoma, and HeLa cell lines (Villarante et al. 2011), without discussing its mode of action.

PYOCINS

Pyocins are produced by *Pseudomonas* sp. The pyocin genes are present on the *P. aeruginosa* chromosome, and their activities are induced by mutagenic agents. There are three types of pyocins: (1) R-type pyocins resemble nonflexible and contractile tails of bacteriophages. These are nuclease and protease resistant and cause depolarization of the cytoplasmic membrane. (2) F-type pyocins resemble phage tails containing flexible and noncontractile rod-like structures. (3) S-type pyocins are colicin-like, protease-sensitive bacteriocins. Abdi-Ali et al. (2004) demonstrated the cytotoxicity of pyocin on tumor cell lines HepG2 and Im9 in a dose-dependent manner. The cytotoxic effect of pyocinS2 was also reported by Watanabe and Saito (1980) against cancerous cell lines (HeLa, AS-II derived from embryonal carcinoma of ovary, simian virus-40 transformed mouse kidney cell line mKS-ATU-7) and normal mice cells (BALB/3T3). However, it showed noncytotoxicity against some of the cancerous (HCG-27) and normal cells.

MICROCINS

Microcins are small sized (<10 kDa) bacteriocins secreted by *Enterobacteria* (mostly *E. coli*) under conditions of nutrient depletion. To date, 14 microcins have been reported, out of which only seven have been isolated and characterized. The toxicity of microcin E492 has been demonstrated against various human malignant cell lines such as HeLa (human cervical adenocarcinoma), Jurkat (T-cell derived from acute T-cell leukemia), RJ2.25 (a variant of Burkitt's lymphoma), and colorectal carcinoma cells. The peptide showed no effect against normal bone marrow cells, splenocytes, KG-1, and nontumor macrophage-derived cells (Hetz et al. 2002). It was also observed that M-E492 caused apoptosis and necrosis at lower and higher concentrations, respectively, showing cell shrinkage, DNA fragmentation and extracellular exposure of phosphatidylserine, loss of mitochondrial membrane potential, and release of calcium ions from intracellular stores (Hetz et al. 2002). The *in vivo* effects of M-E492 on human tumor cells were studied in a preclinical model using mice, showing that M-E492 fibrils administered had anticancer activity (Lagos et al. 2009).

COLICINS

Colicins are plasmid-encoded high molecular weight (>20 kDa) peptides secreted by *Enterobacteriaceae*. Colicins show anticancer activities against several human tumor cell lines such as breast cancer, colon cancer, bone cancer, and uteri cell line HeLa. Chumchalova and Smarda (2003) demonstrated the effects of four pure colicins—A, E1, U, and E3—on 11 human cancer cell lines using MTT (tetrazolium-bromide) assay, showing colicin E1 and A to be the most potent. In another report, colicin E3 exhibited complete killing of human uteri carcinoma cell line (Smarda et al. 1978). Colicin E3 also showed time- and dose-dependent anticancer activity against murine leukemia cells P388 (Fuska et al. 1978). The direct administration of colicin E3 into the subcutaneous nodes of solid HK-adenocarcinoma showed significant reduction in mean mass of tumor in mice (Cursino et al. 2002). The mode of action of colicins is not yet known, but a few studies reported that colicin E3 cleaves the 18S rRNA of the isolated eukaryotic ribosomes (Turnowsky et al. 1973; Suzuki 1978). Colicins kill cancer cells by creating pores in the plasma membrane, thereby activating apoptosis (Chumchalova and Smarda 2003).

BACTERIA-MEDIATED CANCER THERAPY—PROS AND CONS

In the current scenario, the emerging resistance to conventional anticancer therapies in patients with advanced solid tumors has prompted the urgency of alternative cancer therapies. In this regard, live, attenuated, or genetically modified nonpathogenic bacterial species have shown a broad capability to multiply selectively in cancer cells and inhibit their growth. These bacteria and their spores served as ideal vectors in order to deliver therapeutic proteins to specific cancer cells. In addition to this, bacterial toxins have also emerged as promising anticancer agents. These novel outcomes may be a solution to the increasing incidence of cancer deaths and the limitations of high-cost conventional treatments. Most importantly, the bacteria-mediated cancer treatment is cheap, safe, and very effective. Bacteria can grow preferentially in hypoxic areas of cancer cells, where rarely is blood perfusion enough for carrying currently available anticancer drugs. The bioactive peptides or bacteriocins have also shown selective cytotoxicity toward tumors as compared to normal cells, thus making them promising anticancer agents for clinical trials studies. Bacteriocins are not only nonimmunogenic but are also biodegradable in nature.

Despite the broad-spectrum anticancer activities of bacteria and their bioactive peptides, drawbacks of this strategy are also known to us. Although bacteria have shown significant *in vivo* results, further investigation regarding their mode of action against specific cancer cells is required to make them complete therapeutic agents. In order to utilize *Salmonella* sp. as potential anticancer agents, more intense research is still required for disclosing the optimal protocols to explore this natural anticancer agent. The mode of administration is the major drawback of protein-mediated drugs. One of the major challenges for the use of bacteriocins as anticancer drugs is to improve the stability of bacteriocins. The production and purification of bacteriocins on a large scale are additional hurdles because the typical concentrations of purified bacteriocins are less than a milligram per liter of the bacterial culture. Furthermore, the delivery of peptide-derived anticancer drugs is another challenge due to larger size in comparison with the commercially available antibiotics. In addition to this, the hydrophobicity, charge, and biocompatibility of peptides may also be a hurdle to the development of efficacious anticancer drugs.

CONCLUDING REMARKS AND FUTURE PERSPECTIVES

In view of the millions of deaths per year due to cancer, several bacterial species have been investigated so far as cancer treatment modalities. Live, attenuated bacteria as anticancer agents and vectors for gene-directed enzyme prodrug therapy have been studied as potential candidates. Chimeric toxins are also being investigated as toxin-based anticancer therapies. Moreover, *S. typhimurium* showed a significant role as a very successful vector for cytotoxic agents, making bacteria very promising candidates. Some bacteriocins have also shown *in vitro* cytotoxicity against various cancer cells at lower concentrations.

In spite of successful outcomes from bacteria-mediated cancer therapy, the successful translation of these preclinical reports into clinical practice will depend on the clinical trials strategies. In fact, because cancer is a multifactorial disease, a combination of recombinant DNA technology along with immunotherapy applied to the bacteria will serve as the foundation for cancer treatment in the future. Further, as bacteriocins are nonimmunogenic and biodegradable, their potentiality to serve as synergistic agents to conventional anticancer drugs needs to be tested. Finally, the development and marketing of bacteria-mediated potential anticancer drugs requires not only hard work but also marketable commodities that help generate wealth both for the research institutes and the country.

ACKNOWLEDGMENT

This study was partially supported by Maulana Azad National Fellowship (F1-17.1/2015-16/MANF-2015-17-BIH-60730), University Grants Commission, Delhi, India.

REFERENCES

Abdi-Ali A, Worobec EA, Deezagi A, Malekzadeh F. Cytotoxic effects of pyocin S2 produced by *Pseudomonas aeruginosa* on the growth of three human cell lines. *Can J Microbiol*. 2004;50:375–381.

Agrawal N, Bettegowda C, Cheong I, Geschwind J, Drake CG, Hipkiss EL, et al. Bacteriolytic therapy can generate a potent immune response against experimental tumors. *Proc Natl Acad Sci USA*. 2004;101:15172–15177.

Ahsan A, Ray D, Ramanand SG, Hegde A, Whitehead C, Rehemtulla A, et al. Destabilization of the epidermal growth factor receptor (EGFR) by a peptide that inhibits EGFR binding to heat shock protein 90 and receptor dimerization. *J Biol Chem*. 2013;288:26879–26886.

Ansiaux R, Gallez B. Use of botulinum toxins in cancer therapy. *Expert Opin Investig Drugs*. 2007;16:209–218.

Basu D, Herlyn M. *Salmonella typhimurium* as a novel RNA interference vector for cancer gene therapy. *Cancer Biol Ther*. 2008;7:1–2.

Bauer H, Darji A, Chakraborty T, Weiss S. *Salmonella*-mediated oral DNA vaccination using stabilized eukaryotic expression plasmids. *Gene Ther*. 2005;12:364–372.

Beaulieu L. *Production, Purification et Caracterisation de la Pediocine PA-1 Naturelle et de ses Formes Recombiantes: Contribution a la Mise en Evidence d' une Nouvelle Activite Biologique*. Quebec, Canada: Universite Laval; 2004.

Bernardes N, Ribeiro AS, Abreu S, Mota B, Matos RG, Arraiano CM, et al. The bacterial protein azurin impairs invasion and FAK/Src signaling in P-cadherin over expressing breast cancer cell models. *PLoS One*. 2013;8:e69023.

Bernardes N, Ribeiro AS, Abreu S, Vieira AF, Carreto L, Santos M, et al. High throughput molecular profiling of a P-cadherin overexpressing breast cancer model reveals new targets for the anti-cancer bacterial protein azurin. *Int J Biochem Cell Biol*. 2014;50:1–9.

Bizzarri AR, Santini S, Coppari E, Bucciantini M, Di Agostino S, Yamada T, et al. Interaction of an anticancer peptide fragment of azurin with p53 and its isolated domains studied by atomic force spectroscopy. *Int J Nanomedicine*. 2011;6:3011–3019.

Carey R, Holland J, Whang H, Neter E, Bryant B. Clostridial oncolysis in man. *Eur J Cancer*. 1967;3:37–46.

Carswell EA, Old LJ, Kassel RL, Green S, Fiore N, Williamson B. An endotoxin-induced serum factor that causes necrosis of tumors. *Proc Natl Acad Sci*. 1975;72:3666–3670.

Chan SC, Hui L, Chen HM. Enhancement of the cytolytic effect of anti-bacterial cecropin by the microvilli of cancer cells. *Anticancer Res*. 1998;18:4467–4474.

Chanana V, Majumdar S, Rishi P. Involvement of caspase-3, lipid peroxidation and TNF-alpha in causing apoptosis of macrophages by coordinately expressed *Salmonella* phenotype under stress conditions. *Mol Immunol*. 2007;44:1551–1558.

Chander H, Majumdar S, Sapru S, Rishi P. 55 kDa outer-membrane protein from short-chain fatty acids exposed *Salmonella enterica* serovar typhi induces apoptosis in macrophages. *Antonie Van Leeuwenhoek*. 2006;89:317–323.

Chumchalova J, Smarda J. Human tumor cells are selectively inhibited by colicins. *Folia Microbiol*. 2003;48:111–115.

Coley WB. The treatment of malignant tumors by repeated inoculations of erysipelas. With a report often original cases. *Clin Orthop Relat Res*. 1991;1893:3–11.

Cursino L, Smarda J, Chartone-Souza E, Nascimento AMA. Recent updated aspects of colicins of Enterobacteriaceae. *Braz J Microbiol.* 2002;33:185–195.

Diep DB, Havarstein LS, Nes IF. Characterization of the locus responsible for the bacteriocin production in *Lactobacillus plantarum* C11. *J Bacteriol.* 1996;178:4472–4483.

Falnes PO, Ariansen S, Sandwig K, Olsnes S. Requirement for prolonged action in the cytosol for optimal protein synthesis inhibition by diphtheria toxin. *J Biol Chem.* 2000;275:4363–4368.

Fensterle J, Bergmann B, Yone CL, Hotz C, Meyer SR, Spreng S, et al. Cancer immunotherapy based on recombinant *Salmonella enterica serovar typhimurium* aroA strains secreting prostate-specific antigen and cholera toxin subunit B. *Cancer Gene Ther.* 2008;15:85–93.

Ferlay J, Soerjomataram I, Ervik M, Dikshit R, Eser S, Mathers C, et al. *GLOBOCAN 2012 v1.0, Cancer Incidence and Mortality Worldwide: IARC Cancer Base No. 11* [Internet]. Lyon, France: International Agency for Research on Cancer; 2013.

Fialho AM, Stevens FJ, Das Gupta TK, Chakrabarty AM. Beyond host-pathogen interactions: Microbial defense strategy in the host environment. *Curr Opin Biotechnol.* 2007;18:279–286.

Frankel AE, Rossi P, Kuzel TM, Foss F. Diphtheria fusion protein therapy of chemoresistant malignancies. *Curr Cancer Drug Targets.* 2002;2:19–36.

Fu W, Lan H, Li S, Han X, Gao T, Ren D. Synergistic antitumor efficacy of suicide/ePNP gene and 6-methylpurine 2′- deoxyriboside via *Salmonella* against murine tumors. *Cancer Gene Ther.* 2008;15:474–484.

Fujimori M, Amano J, Taniguchi S. The genus Bifidobacterium for cancer gene therapy. *Curr Opin Drug Discov Devel.* 2003;5:200–203.

Fuska J, Fuskova A, Smarda J, Mach J. Effect of colicin E3 on leukemia cells P388 *in vitro.* *Experientia.* 1978;35:406–407.

Galan JE. *Salmonella* interactions with host cells: Type III secretion at work. *Annu Rev Cell Dev Biol.* 2001;17:53–86.

Goel HL, Pursell B, Standley C, Fogarty K, Mercurio AM. Neuropilin-2 regulates $\alpha 6\beta 1$ integrin in the formation of focal adhesions and signaling. *J Cell Sci.* 2012;125:497–506.

Greenfield L, Johnson VG, Youle RJ. Mutations in diphtheria toxin separate binding from entry and amplify immunotoxin selectivity. *Science.* 1987;238:536–539.

Guiney DG. The role of host cell death in *Salmonella* infections. *CTMI.* 2005;289:131–150.

Hanahan D, Weinberg RA. The hallmarks of cancer: The next generation. *Cell.* 2011;144:646–673.

Hansen JN, Liu W. Some chemical and physical properties of nisin, a small-protein antibiotic produced by *Lactococcus lactis.* *Appl Environ Microbiol.* 1990;56:2551–2558.

Hetz C, Bono MR, Barros LF, Lagos R. Microcin E492, a channel-forming bacteriocin from *Klebsiella pneumoniae*, induces apoptosis in some human cell lines. *Proc Natl Acad Sci USA.* 2002;99:2696–2701.

Hidaka A, Hamaji Y, Sasaki T, Taniguchi S, Fujimori M. Exogenous cytosine deaminase gene expression in *Bifidobacterium breve* I-53-8w for tumor-targeting enzyme/prodrug therapy. *Biosci Biotechnol Biochem.* 2007;71:2921–2926.

Hoskin DW, Ramamoorthy A. Studies on anticancer activities of antimicrobial peptides. *Biochem Biophys Acta* 2008;1778:357–375.

Joo NE, Ritchie K, Kamarajan P, Miao D, Kapila YI. Nisin, an apoptogenic bacteriocin and food preservative, attenuates HNSCC tumorigenesis via CHAC1. *Cancer Med.* 2012;1:295–305.

Kamarajan P, Hayami H, Matte B, Liu Y, Danciu T, Ramamoorthy A, et al. Nisin ZP, a bacteriocin and food preservative, inhibits head and neck cancer tumorigenesis and prolongs survival. *PLoS One.* 2015;10:e0131008.

Karsten V, Pike J, Troy K, Luo X, Zheng LM, King I, et al. A strain of *Salmonella typhimurium* VNP20009 expressing an anti-angiogenic peptide from platelet factor-4 has enhanced anti-tumor activity. *Proc Annu Meet Am Assoc Cancer Res*. 2001;42:3700.

Kaspar AA, Reichert JM. Future directions for peptide therapeutics development. *Drug Discov Today*. 2013;18:807–817.

Klaenhammer TR, Fremaux C, Ahn C, Milton K. Molecular biology of bacteriocins produced by *Lactobacillus*. In: Hoover DG, Steenson LR, eds. *Bacteriocins of Lactic Acid Bacteria*. New York, NY: Academic Press; 1993:151–180.

Knodler LA, Steele-Mortimer O. Taking possession: Biogenesis of the *Salmonella*-containing vacuole. *Traffic*. 2003;4:587–599.

Kokai Kun JF, Mcclane BA. Determination of functional regions of *Clostridium perfringens* enterotoxin through deletion analysis. *Clin Infect Dis*. 1997;25:S165–S167.

Kokai Kun JF, Benton K, Wieckowski EU, Mcclane BA. Identification of a *Clostridium perfringens* enterotoxin region required for large complex formation and cytotoxicity by random mutagenesis. *Infect Immun*. 1999;67:5634–5641.

Kumar B, Kaur B, Balgir PP, Garg N. Cloning and expression of bacteriocins of *Pediococcus* spp. *Arch Clin Microbiol*. 2011;2:1–18.

Kumar B, Balgir PP, Kaur B, Mittu B, Chauhan A. *In vitro* cytotoxicity of native and rec-pediocin cp2 against cancer cell lines: A comparative study. *Open Access Sci Rep*. 2012;1:316–321.

Lagos R, Tello M, Mercado G, Garcia V, Monasterio O. Antibacterial and antitumorigenic properties of microcin E492, a pore-forming bacteriocin. *Curr Pharm Biotechnol*. 2009;10:74–85.

Lanzerin M, Sand O, Olsnes S. GPI-anchored diphtheria toxin receptor allows membrane translocation of the toxin without detectable ion channel activity. *EMBO J*. 1996;15:725–734.

Leschner S, Weiss S. Salmonella—Allies in the fight against cancer. *J Mol Med*. 2010;88:763–773.

Li X, Fu GF, Fan YR, Liu WH, Liu XJ. *Bifidobacterium adolescentis* as a delivery system of endostatin for cancer gene therapy: Selective inhibitor of angiogenesis and hypoxic tumor growth. *Cancer Gene Ther*. 2003;10:105–111.

Lin SL, Spinka TL, Le TX, Pianta TJM, King I, Belcourt MF, et al. Tumor directed delivery and amplification of tumor-necrosis factor-α (TNF) by attenuated *Salmonella typhimurium*. *Clinical Cancer Res*. 1999;5:3822.

Louie GV, Yang W, Bowman ME, Choe S. Crystal structure of the complex of diphtheria toxin with an extracellular fragment of its receptor. *Mol Cell*. 1997;1:67–68.

Low KB, Ittensohn M, Lin S, Clairmont C, Luo X, Zheng LM, et al. VNP20009, a genetically modified *Salmonella typhimurium* for treatment of solid tumors. *Proc Am Assoc Cancer Res*. 1999;40:851.

Luo M, Guan JL. Focal adhesion kinase: A prominent determinant in breast cancer initiation, progression and metastasis. *Cancer Lett*. 2010;289:127–139.

Luo X, Ittensohn M, Low B, Pawelek J, Li Z, Ma X, et al. Genetically modified *Salmonella typhimurium* inhibited growth of primary tumors and metastase. *Proc Annu Meet Am Assoc Cancer Res*. 1999;40:3146.

Luo X, Li Z, Shen SY, Runyan JD, Bermudes D, Zheng LM. Genetically armed *Salmonella typhimurium* delivered therapeutic gene and inhibited tumor growth in preclinical models. *Proc Annu Meet Am Assoc Cancer Res*; 2001;42:3693.

Maher S, McClean S. Investigation of the cytotoxicity of eukaryotic and prokaryotic antimicrobial peptides in intestinal epithelial cells *in vitro*. *Biochem Pharmacol*. 2006;71:1289–1298.

Majno G, Joris I. Apoptosis, oncosis, and necrosis. An overview of cell death. *Am J Pathol*. 1995;146:3–15.

Marsden VS, O'Connor L, O'Reilly LA, Silke J, Metcalf D, Ekert PG, et al. Apoptosis initiated by Bcl-2- regulated caspase activation independently of the cytochrome c/Apaf-1/caspase-9 apoptosome. *Nature*. 2002;419:634–637.

Micewicz ED, Jung CL, Schaue D, Luong H, McBride WH, Ruchala P. Small azurin derived peptide targets ephrin receptors for radiotherapy. *Int J Pept Res Ther*. 2011;17:247–257.

Milletti F. Cell-penetrating peptides: Classes, origin, and current landscape. *Drug Discov Today*. 2012;17:850–860.

Minton NP. Clostridia in cancer therapy. *Nat Rev Microbiol*. 2003;1:237–242.

Moreno M, Kramer MG, Yim L, Chabalgoity JA. *Salmonella* as live Trojan horse for vaccine development and cancer gene therapy. *Curr Gene Ther*. 2010;10:56–76.

Nagakura C, Hayashi K, Zhao M, Yamauchi K, Yamamoto N, Tsuchiya H, et al. Efficacy of a genetically-modified *Salmonella typhimurium* in an orthotopic human pancreatic cancer in nude mice. *Anticancer Res*. 2009;29:1873–1878.

Nemunaitis J, Cunningham C, Senzer N, Kuhn J, Cramm J, Litz C, et al. Pilot trial of genetically modified, attenuated *Salmonella* expressing the *E. coli* cytosine deaminase gene in refractory cancer patients. *Cancer Gene Ther*. 2003;10:737–744.

Nougayrede JP, Taieb F, De Rycke J, Oswald E. Cyclomodulins: Bacterial effectors that modulate the eukaryotic cell cycle. *Trends Microbiol*. 2005;13:103–110.

Ohl ME, Miller SI. *Salmonella*: A model for bacterial pathogenesis. *Annu Rev Med*. 2001;52:259–274.

Paesold G, Guiney DG, Eckmann L, Kagnoff MF. Genes in the *Salmonella* pathogenicity island 2 and the *Salmonella* virulence plasmid are essential for Salmonella-induced apoptosis in intestinal epithelial cells. *Cell Microbiol*. 2002;4:771–781.

Paiva AD, Breukink E, Mantovani HC. Role of lipid II and membrane thickness in the mechanism of action of the lantibiotic bovicin HC5. *Antimicrob Agents Chemother*. 2011;55:5284–5293.

Paiva AD, deOliveira MD, dePaula SO, Baracat-Pereira MC, Breukink E, Mantovani HC. Toxicity of bovicin HC5 against mammalian cell lines and the role of cholesterol in bacteriocin activity. *Microbiology*. 2012;158:2851–2858.

Papagianni M. Ribosomally synthesized peptides with antimicrobial properties: Biosynthesis, structure, function, and applications. *Biotechnol Adv*. 2003;21:465–499.

Pawelek JM, Low KB, Bermudes D. Tumor targeted *Salmonella* as a novel anticancer vector. *Cancer Res*. 1997;57:4537–4544.

Pawelek JM, Low KB, Bermudes D. Bacteria as tumour targeting vectors. *Lancet Oncol*. 2003;4:548–556.

Preet S, Bharati S, Panjeta A, Tewari R, Rishi P. Effect of nisin and doxorubicin on DMBA-induced skin carcinogenesis—A possible adjunct therapy. *Tumor Biol*. 2015;36(11):8301–8308.

Puri RK. Development of a recombinant interleukin-4-Pseudomonas exotoxin for therapy of glioblastoma. *Toxicol Pathol*. 1999;27:53–57.

Riedl S, Rinner B, Asslaber M, Schaider H, Walzer S, Novak A, et al. In search of a novel target—Phosphatidylserine exposed by non-apoptotic tumor cells and metastases of malignancies with poor treatment efficacy. *Biochim Biophys Acta*. 2011;1808:2638–2645.

Saito H, Watanabe T. Effect of a bacteriocin produced by *Mycobacterium smegmatis* on growth of cultured tumor and normal cells. *Cancer Res*. 1979;39:5114–5117.

Saito H, Watanabe T. Effects of a bacteriocin from *Mycobacterium smegmatis* on BALB/3T3 and Simian virus 40–transformed BALB/c mouse cells. *Microbiol Immunol*. 1981;25:13–22.

Saito H, Watanabe T, Tomioka H. Purification, properties and cytotoxic effect of a bacteriocin from *Mycobacterium smegmatis*. *Antimicrob Agents Chemother*. 1979;15:504–509.

Saltzman DA, Heise CP, Hasz DE, Zebede M, Kelly SM, Curtiss R. Attenuated *Salmonella typhimurium* containing interleukin-2 decreases MC-38 hepatic metastases: A novel anti-tumor agent. *Cancer Biother Radiopharm.* 1996;11:145–153.

Saltzman DA, Katsanis E, Heise CP, Hasz DE, Kelly SM, Curtiss R. Patterns of hepatic and splenic colonization by an attenuated strain of *Salmonella typhimurium* containing the gene for human interleukin-2: A novel anti-tumor agent. *Cancer Biother Radiopharm.* 1997;12:37–45.

Sand SL, Haug TM, Nissen-Meyer J, Sand O. The bacterial peptide pheromone plantaricin A permeabilizes cancerous, but not normal, rat pituitary cells and differentiates between the outer and inner membrane leaflet. *J Membrane Biol.* 2007;216:61–71.

Sand SL, Oppegård C, Ohara S, Iijima T, Naderi S, Blomhoff HK, et al. Plantaricin A, a peptide pheromone produced by *Lactobacillus plantarum*, permeabilizes the cell membrane of both normal and cancerous lymphocytes and neuronal cells. *Peptides.* 2010;31:1237–1244.

Sand SL, Nissen-Meyer J, Sand O, Haug TM. Plantaricin A, a cationic peptide produced by *Lactobacillus plantarum*, permeabilizes eukaryotic cell membranes by a mechanism dependent on negative surface charge linked to glycosylated membrane proteins. *Biochim Biophys Acta.* 2013;1828:249–259.

Smarda J, Obdrzalek V, Taborsky I, Mach J. The cytotoxic and cytocidal effect of colicin E3 on mammalian tissue cells. *Folia Microbiol.* 1978;23:272–277.

Sok M, Sentjurc M, Schara M. Membrane fluidity characteristics of human lung cancer. *Cancer Lett.* 1999;139:215–220.

Stritzker J, Weibel S, Hill PJ, Oelschlaeger TA, Goebel W, Szalay AA. Tumor-specific colonization, tissue distribution, and gene induction by probiotic *Escherichia coli* Nissle 1917 in live mice. *Int J Med Microbiol.* 2007;297:151–162.

Suzuki H. Colicin E3 inhibits rabbit globin synthesis. *FEBS Lett.* 1978;89:121–125.

Takaya A, Suzuki A, Kikuchi Y, Eguchi M, Isogai E, Tomoyasu T, et al. Derepression of *Salmonella* pathogenicity island 1 genes within macrophages leads to rapid apoptosis via caspase-1- and caspase-3-dependent pathways. *Cell Microbiol.* 2005;7:79–90.

Taylor BN, Mehta RR, Yamada T, Lekmine F, Christov K, Chakrabarty AM, et al. Noncationic peptides obtained from azurin preferentially enter cancer cells. *Cancer Res.* 2009;69:537–546.

Theys J, Landuyt W, Nuyts S, Van Mellaert L, van Oosterom A, Lambin P, et al. Specific targeting of cytosine deaminas to solid tumors by engineered *Clostridium acetobutylicum. Cancer Gene Ther.* 2001;8:294–297.

Thiele E, Arison R, Boxer G. Oncolysis by Clostridia IV effect of nonpathogenic Clostridial spores in normal and pathological tissues. *Cancer Res.* 1963;24:234–238.

Turnowsky F, Drews J, Eich F, Hogenauer G. In vitro inactivation of ascites ribosomes by colicin E3. *Biochem Biophys Res Commun.* 1973;52:327–334.

Vazquez-Torres A, Jones-Carson J, Bäumler AJ, Falkow S, Valdivia R, Brown W, et al. Extraintestinal dissemination of *Salmonella* by CD18-expressing phagocytes. *Nature.* 1999;401:804–808.

Vendrell A, Gravisaco MJ, Pasetti MF, Croci M, Colombo L, Rodriguez C, et al. A novel *Salmonella typhi*-based immunotherapy promotes tumor killing via an antitumor Th1-type cellular immune response and neutrophil activation in a mouse model of breast cancer. *Vaccine.* 2010;29:728–736.

Villarante KI, Elegado FB, Iwatani S, Zendo T, Sonomoto K, de Guzman EE. Purification and characterization and *in vitro* cytotoxicity of the bacteriocin from *Pediococcus acidilactici* K2a2-3 against human colon adenocarcinoma (HT29) and human cervical carcinoma (HeLa) cells. *World J Microbiol Biotechnol.* 2011;27:975–980.

Watanabe T, Saito H. Cytotoxicity of pyocin S2 to tumor and normal cells and its interaction with cell surfaces. *Biochim Biophys Acta.* 1980;633:77–86.

Weinrauch Y, Zychlinsky A. The induction of apoptosis by bacterial pathogens. *Annu Rev Microbiol.* 1999;53:155–187.

Yamada T, Hiraoka Y, Ikehata M, Kimbara K, Avner BS, Das Gupta TK, et al. Apoptosis or growth arrest: Modulation of tumor suppressor p53's specificity by bacterial redox protein azurin. *Proc Natl Acad Sci USA.* 2004;101:4770–4775.

Yamada T, Mehta RR, Lekmine F, Christov K, King ML, Majumdar D, et al. A peptide fragment of azurin induces a p53-mediated cell cycle arrest in human breast cancer cells. *Mol Cancer Ther.* 2009;8:2947–2958.

Yoon WS, Chae YS, Hong J, Park YK. Antitumor therapeutic effects of a genetically engineered *Salmonella typhimurium* harbouring TNF-alpha in mice. *Appl Microbiol Biotechnol.* 2011;89:1807–1819.

Yu YA, Zhang Q, Szalay AA. Establishment and characterization of conditions required for tumor colonization by intravenously delivered bacteria. *Biotechnol Bioeng.* 2008;100:567–578.

Zhao M, Yang M, Li XM, Jiang P, Baranov E, Li S, et al. Tumor targeting bacterial therapy with amino acid auxotrophs of GFP-expressing *Salmonella typhimurium*. *Proc Natl Acad Sci USA.* 2005;102:755–760.

Zhao H, Sood R, Jutila A, Bose S, Fimland G, Nissen-Meyer J, et al. Interaction of the antimicrobial peptide pheromone plantaricin A with model membranes: Implications for a novel mechanism of action. *Biochim Biophys Acta.* 2006;1758:1461–1474.

Zhao M, Geller J, Ma H, Yang M, Penman S, Hoffman RM. Monotherapy with a tumor-targeting mutant of *Salmonella typhimurium* cures orthotopic metastatic mouse models of human prostate cancer. *Proc Natl Acad Sci USA.* 2007;104:10170–10174.

6 Immunotherapy
Improving Survival, Quality of Life in Rapidly Progressing Head and Neck Cancer

Sowmiya Renjith and Sathya Chandran

CONTENTS

INTRODUCTION

Head and neck squamous cell carcinoma (HNSCC) denotes a heterogeneous disease entity, which comprises a variety of tumors originating in the lip/oral cavity, hypopharynx, oropharynx, nasopharynx, or larynx with differences in epidemiology, etiology, and therapeutical approach. It is the sixth most common malignancy worldwide, accounting for approximately 6% of all cases, and is responsible for an estimated 1%–2% of all cancer deaths (Ferlay et al. 2010).

HNSCC has been associated historically with alcohol and tobacco use; however, in the past decade, infection with high-risk human papillomaviruses (HPVs) and especially type 16 has been implied in the pathogenesis of a subordinate class of HNSCCs, mainly those arising from the oropharynx. HPV-associated oropharyngeal cancer represents a unique biological and clinical entity with a better favorable prognosis (Weinberger et al. 2006; Rampias et al. 2009). This raises the question whether HPV-positive patients should be treated with less intensive treatment, which is currently being addressed in clinical trials. The majority of HNSCC patients present with locally advanced disease for which a multimodality therapeutic approach is employed. For recurrent/metastatic (R/M) disease, cytotoxic-based chemotherapy remains the standard therapeutic option, and the median survival of patients treated with palliative chemotherapy alone ranges from 6 to 10 months (Baxi et al. 2012).

Cancer is a multistep process originating from alterations in genetic normal proliferation and differentiation. In a normally functioning environment, immune surveillance acts as an effective tumor suppressor entity, as the above alterations trigger the development of tumor-related antigens initially recognized by the immune system. However, it is evident that after the equilibrium phase, the immune system might lose the ability to eradicate cancer cells, or new mutations might render tumor cells poorly immunogenic and resistant to elimination by immune cells (Dunn et al. 2004; Raval et al. 2014; Mendes et al. 2016).

Both the self-incorporated and adaptive immune systems have the ability to differentiate between self and non-self pathogens. Innate (self-incorporated) immunity is based on nonspecific defense mechanisms that are activated immediately after contact with a pathogen; limited numbers of receptors are used that are encoded in the germ line and are able to recognize features common to many pathogens. In contrast, adaptive immunity works upon somatic cell gene rearrangements to produce a multitude of antigen receptors that differentiate between closely related molecules; it is mainly moved by highly specific antigen receptors on T and B cells and is highly specific to a particular pathogen (Medzhitov and Janeway 2002).

CANCER DEVELOPMENT AND THE IMMUNE SYSTEM

The immune system is able to identify and eliminate tumor cells based on their association with tumor-associated antigens (TAAs) via a process known as immune surveillance. Tumors can express microbial, mutated, and fusion proteins. The immune system can also recognize aberrantly expressed self-proteins (Vesely et al. 2011). During evolution of a tumor, the T cells are activated on encounters with antigen-presenting cells (APCs), usually dendritic cells (DCs), B cells, or macrophages that display TAAs, which are bound to major histocompatibility complex (MHC) proteins and further interact with T-cell receptors (TCRs). A sophisticated network of co-stimulatory and co-inhibitory pathways, which normally play a vital role in the inhibition of auto-immunity, are manipulated by cancer cells to avoid immune surveillance. Co-stimulatory or activating receptors include CD28, CD137, CD40, and OX-40 (Pardoll 2012). On secretion of specific chemokines, a proportion of T cells differentiate into cytotoxic $CD8^+$ cells that move to the tumor microenvironment and directly attack tumor cells. Using gene expression profiling, many studies initially conducted in malignant melanoma gave way to identification and description of two separate subtypes of tumor microenvironment based on the presence of a transcriptional profile denotative of T-cell infiltration (Woo et al. 2015). More specifically, an "inflamed" tumor immune-phenotype is characterized by the recruitment of T cells, immune signals, and chemokines, whereas a "noninflamed" tumor phenotype lacks instantaneous infiltration of T cells and other immune modulators. Most importantly, it has been suggested that within the subset of patients among "inflamed phenotype," tumor progression might have been a result of negative immune regulators acting at the level of the microenvironment. Controversially, failure in noninflamed tumors is due to poor effector T-cell trafficking at the site of the tumor (Spranger and Gajewski 2013; Gajewski 2015).

Lower survival rates in combination with significant toxicities caused by the present treatment strategies applied in HNSCC emphasize an immediate need for enhanced treatment options. It has been widely accepted that the immune system plays a pivotal role in cancer development, as tumor cells evade immune surveillance by escaping the inhibitory checkpoint pathways that suppress antitumor T-cell responses (Zou and Chen 2008). HNSCC has been heavily studied as an immune-suppressive disease. Following a greater understanding of the underlying mechanisms behind the control of malignancies by the immune system, the evolution of immune-based therapies has emerged as a widely accepted approach for the treatment of cancer.

In this chapter, we discuss the role of the immune system in HNSCC tumorigenesis and describe immunotherapy approaches currently under investigation.

ROLE OF IMMUNE SYSTEM IN HNSCC

Current evidence provides a pivotal role of the immune system in the development and evolution of HNSCC. Further, the level of the immune system is most likely to be of prognostic value in HNSCC. HNSCC is regarded as an immune-suppressive disease, characterized by deregulations of immune-competent cells and impaired cytokine excretion (Varilla et al. 2013). Immunosuppressive patients are more prone to develop head and neck cancer, and prognosis is evidently poor (Schoenfeld 2015). For HPV-negative HNSCC patients, though there is a strong etiologic association with tobacco and alcohol, it is hypothesized that tumor progress concludes the inability of the body's immune system to eliminate the tumor. As numerous cells of the immune system provide a complex network of defense, the balance within subsets of T lymphocytes, in combination with the effects of the tumor microenvironment, was believed to modulate antitumor immunity (Chikamatsu et al. 2007). T cells, macrophages, dendritic cells, and natural killer (NK) cells are unique players in the tumor microenvironment, whose functional alterations modify the immune system response.

Individuals with HNSCC have reduced antitumor immune responses, and tumor progression or regression is believed to be related to immune dysfunction. Many mechanisms, such as the presence of tumor-secreted proteins that act as inhibitory stimuli, cytokines, and T-cell apoptosis have also been referred to contribute to immune deregulation (Whiteside 2004). Effectively, the presence of T-regulatory cells (Tregs) has emerged as a potential mechanism of immunomodulation in HNSCC. Tregs depress or downregulate induction and proliferation of effector T cells and have been consistently observed at a higher frequency in individuals with HNSCC (Hoffmann et al. 2002; Reichert et al. 2002). Interestingly, unlike other solid malignancies such as lung cancer and renal cell cancer, the involvement of Tregs has been correlated with good clinical outcome (Badoual et al. 2006; Siddiqui et al. 2007). A multitude of functions of Tregs explain the current paradox, even including inflammation suppression triggered by immune cells, elimination of macrophages that have a pro-tumorous effect in cancer development, and induction of apoptosis (Badoual et al. 2009). Recent studies suggest the presence of different subsets of Tregs with functional heterogeneity (Sun et al. 2014). Meanwhile, the low levels of CD4[+] and CD8[+] T cells have been found in individuals with HNSCC, exclusively those with active disease, suggesting decreased function of effector cells in this current population. Furthermore, even in patients without any evidence of disease, immune abnormalities might persist after weeks or years of curative therapy (Schaefer et al. 2005).

On evidence of HPV infections, HPV-specific effector T cells are considered responsible for elimination of the virus, and HPV-induced oncogenesis has been regarded to correlate with weak HPV-specific T-cell responses (Van der Burg 2012). Meanwhile, PD1, a protein functioning as an immune checkpoint by shielding the activation of T cells, had been found in tonsilar crypts, and PD1 infiltration of lymphocytes has been found in HPV(+)-OPC. The PD1 pathway must play a vital role in HPV-related OPC oncogenesis (Lyford-Pike et al. 2013). Surprisingly, the PD1-positive infiltrating cells have been involved with more favorable clinical outcome (Badoual et al. 2013; Vasilakopoulou et al. 2013).

To date, only a few studies have analyzed the prognostic impact of tumor-infiltrating lymphocytes (TILs) in correlation with HPV status in HNSCC patients. Many studies have demonstrated an increased population of CD8[+] lymphocytes, Tregs, and increased CD8[+]/Treg ratio in HPV[+] OPC, which are involved with improved prognosis (Näsman et al. 2012; Lukesova et al. 2014).

The tumor stem cell hypothesis supports the notion that "in a heterogenic tumor, a subpopulation of tumor cells (cancer stem cells-CSCs) is capable of initiating and expanding a tumor" (Qian et al. 2015: 12). In HNSCC, it has been proposed that delayed growing CSCs escape usual therapies and regenerate the tumor, accounting for the high rates of recurrence (Krishnamurthy and Nor 2012).

Emerging studies and evidence suggest that the host immune system has the potential to recognize CSCs and provoke an immune response. Previous studies in HNSCC have shown that NK cells may particularly target CSCs (Jewett et al. 2012). Moreover, CSCs interact with the tumor microenvironment. In HPV-related HNSCC, the presence of CSCs is controversial (Qian et al. 2013; Tang et al. 2013; Zhang et al. 2014). Surprisingly, CSCs have been related with radio-resistance and cisplatin resistance in HNSCC (Chen et al. 2010; Bourguignon et al. 2012).

Even when immune-editing might eliminate tumor cells with alterations in their antigenic epitope profile, many immune-resistant types escape from the host's immune system by immune-suppressive molecular and cellular mechanisms. Hence, tumors might avoid termination by the immune system through the outgrowth of tumor cells that can suppress, disrupt, or escape the immune system.

In HNSCC, the dominant mechanism of immune-escape is suppression of tumor antigen presentation. Interception of antigen presentation can occur by

1. Downregulation or loss of tumor human leukocyte antigen (HLA) class I molecules expression
2. Disruption of proteins involved in antigen processing, such as TAP1, LMP2, LMP7
3. Suppression of APC function and maturation (Drake et al. 2006; Vesely et al. 2011)

Large-scale next-generation sequencing of HNSCC has shown many mutations in HLA alleles and APM components, but tumor cells avoid absolute loss of HLA expression, since it leads to recognition by NK cells (Ferris 2015). Another hypothesis states that the signal transducer and activator of transcription (STAT) family of proteins modulates the defects in APM, as well as APC function. For example, the deficiency of activated STAT-1 results in decreased expression of APM components, where aberrant signaling of STAT-3 involves impaired tumor antigen presentation by DCs (Wang et al. 2004; Leibowitz et al. 2011). Nevertheless, the disruption of TAA presentation derives from functionally defective circulating lymphocytes that enhanced the levels of apoptosis and increased inhibition induced by Tregs (Lalami and Awada 2016).

Second, the tumor microenvironment contains various immune-suppressive factors from various sources that tumors use as defense mechanisms against the immune system. The HNSCC microenvironment is distinguished by an unarranged cytokine profile, initiating the secretion of immunosuppression over stimulating cytokines. Vascular endothelial growth factor (VEGF), an inflammatory cytokine released from tumor cells, suppresses the DC function; it has been found in high concentrations in patients with HNSCC and has been correlated with regression (Allen et al. 2007; Gildener-Leapman et al. 2013). Upregulation of IL-6 activates the STAT-3 pathway, which adjacently inhibits DC maturation and NK cell, T-cell, and macrophage activation. Higher expression of STAT-3 also increases the production of IL-10 and TGF-b (Yu et al. 2009).

TGF-β has inhibitory effects on effector cells and APCs, whereas IL-10 downregulates the expression of co-stimulatory molecules and MHC (Sato et al. 2011). Moreover, overexpression of diverse chemokines by the HNSCC tumor cells controls the recruitment of suppressive myeloid cells, such as myeloid-derived suppressor cells (MDSCs) and tumor-associated macrophages (TAMs); deregulation of immune cell recruitment inhibits the antitumor immune response for good (Sato et al. 2011; Ferris 2015).

Cancers can modulate through a variety of mechanisms, which are not very well understood, to suppress or prevent attack by immune cells. The most distinguishable example is through expression of PD-L1. *Immune checkpoints*, such as the PD1 pathway, are part of a protein-ligand receptor system that modulates T-cell activation. They are vital for protecting self-tolerance and controlling the duration and amplitude of physiologic immune response, but are changed by the tumor to initiate tumor growth that is hidden by the immune system. In a normal cell, when PD-L1 or PD-L2 combines with PD1, the T cell becomes inactive. This is one of the ways that the host regulates the immune system to avoid an overreaction. Nevertheless, PD-L1 is also exhibited in HNSCC, leading to the disarming of T cells after binding to PD1 on T cells (Tsushima et al. 2006). The concentration of TILs is contrast in association with PD-L1 expression of tumor cells (Cho et al. 2011). CTLA-4 is another immune checkpoint located on the surface of activated CTLs that combines to the B7 ligands found on APCs. T cells have a CD28 receptor that represents a stimulatory counterpart to CTLA4, causing T-cell activation. CTLA-4 competes with the CD28 receptor for binding to the B7 ligand, leading to either an inhibitory or stimulatory effect on T cells (Pardoll 2012).

IMMUNOTHERAPEUTIC STRATEGIES FOR HNSCC

Enhanced understanding of the role of the immune system in cancer has led to the identification of a variety of novel therapeutic targets. Immune-oncology is a currently evolving field of investigation that includes active immunotherapies that are designed to target and harness the patient's own immune system directly to fight a tumor. Even more specifically, it is designed to leverage the unique properties of the immune system (specificity, adaptability, and memory). The intended goal of immunotherapy is to tip the balance in favor of an immune response against the tumor, allowing for tumor eradication or long-term inhibition of tumor growth, and the generation of immunological memory.

MONOCLONAL ANTIBODIES

Cetuximab is a chimeric immunoglobulin G1 (IgG1) monoclonal antibody that has been approved by the U.S. Food and Drug Administration (FDA) in combination with chemotherapy as the standard first-line treatment for R/M HNSCC. It is also used along with radiation for locally advanced HNSCC (Bonner et al. 2006; Vermorken et al. 2008). Cetuximab efficacy is controlled by antibody-dependent cell-mediated cytotoxicity (ADCC), a mechanism of cell-mediated immune defense where NK cells actively destroy a target cell, whose membrane-surface antigen has been bound by cetuximab. NK cells are initiated upon binding to surface receptor FCγRIIIa (Seidel et al. 2013). Moreover, cetuximab triggers a CTL antitumor response through cross-priming of DCs and NKs (Gildener-Leapman et al. 2013). Other anti-EGFR monoclonal antibodies currently increased in HNSCC include panitumumab, nimotuzumab, and zalutumumab. Among them, panitumumab has produced merge results when added to platinum-based chemotherapy in patients with R/M HNSCC (Ermorken et al. 2013). Zalutumumab has demonstrated an OS of 5.3 months and a PFS of 2.1 when administered as monotherapy in individuals with platinum refractory R/M HNSCC (Saloura et al. 2014). Finally, nimotuzumab in combination with (chemo) radiation in locally advanced HNSCC has shown a survival benefit in tumors overexpressing EGFR (Basavaraj et al. 2010).

IMMUNE CHECKPOINT INHIBITORS

It is now evident that tumors co-opt certain immune-checkpoint pathways as a major mechanism of immune resistance, particularly against T cells that are specific for tumor antigens. Because many of the immune checkpoints are triggered by ligand-receptor interactions, they can be actively blocked by antibodies or manipulated by recombinant forms of ligands or receptors. Ipilimumab, a mAb against CTLA-4 that has acquired FDA approval for metastatic melanoma, is presently being evaluated in clinical trials along with cetuximab and intensity-modulated radiation therapy (IMRT) in individuals with advanced HNSCC (NCT01860430 and NCT01935921). A phase 1, open-label, dose escalation study of MGA271 (enoblituzumab, a humanized mAb against CD276 [B7-H3] in combination with ipilimumab in patients with B7-H3–expressing HNSCC and other solid tumors) is also proceeding

(NCT02381314). Tremelimumab is another anti-CTLA4 antibody presently being assessed in clinical trials.

PD1 reacts with two ligands; PD-L1 that is expressed particularly on tumor cells and other immune cells and PD-L2 that is expressed exclusively on macrophages and DCs. Anti-PD1 antibody pembrolizumab (MK-3475) has expressed promising results in HNSCC. Preliminary results from KEYNOTE-012, a phase I study assessing the efficacy of pembrolizumab in patients with R/M HNSCC, had shown a response rate of 20% in PD-L1–positive tumors. Surprisingly, 78% were found to be PD-L1 positive. Reactions were observed in both HPV+ and HPV– patients, but overall survival was better in HPV+ patients. Response duration ranged from 8 to 41 weeks. PD-L1 expression was positively associated with ORR ($p = 0.018$) and PFS ($p = 0.024$). A larger HNSCC expansion cohort of KEYNOTE-012 was recently presented in the 2015 ASCO meeting, expressing an overall response rate (ORR) of 18.2%, whereas 31.3% of patients had stable disease; response rates were similar in HPV+ and HPV– HNSCC. Toxicity was mostly tolerable, with 7.6% of patients experiencing Grade 3 drug-related events (Seiwert et al. 2015). Pembrolizumab is further examined in a multitude of clinical settings in HNSCC. In KEYNOTE-048, it is valuated either as monotherapy or in combination with chemotherapy versus conventional chemotherapy in patients with R/M HNSCC in the initial-line setting (NCT02358031). In KEYNOTE-040, pembrolizumab is taken for comparison to chemotherapy or cetuximab in the second-line setting (NCT02252042). Pembrolizumab is also presently being evaluated in combination with re-irradiation and as part of primary treatment in various clinical trials (NCT02289209, NCT02296684). Anti-PD1 Abs nivolumab (NCT02105636; $n = 340$ patients) and pembrolizumab (NCT02358031; $n = 750$ patients) are being examined as a single agents in randomized phase III trials for platinum-refractory HNSCC. More accurately, a checkmate-141 phase III trial assessed the efficiency of nivolumab versus physician's choice (cetuximab, methotrexate, or docetaxel) in platinum refractory disease. The study was stopped early after an independent monitoring panel determined the primary endpoint of improvement in OS was met with nivolumab. Another promising anti-PD-L1 antibody is durvalumab (MEDI4736), which has shown satisfactory results (~14% response rate as per RECIST criteria, with 24% response rate in PD-L1+ patients) in a phase I trial (Fury et al. 2014). A phase III trial evaluating durvalumab alone or in combination with tremelimumab compared to conventional treatment is proceeding in patients with advanced HNSCC (NCT02369874).

Another group of receptors with a manipulating effect on immune cells includes other checkpoint receptors such as LAG-3 or the killer-cell immunoglobulin-like receptors (KIRs) (Campbell and Purdy 2011). They regulate immune response by interaction with MHC I molecules. Most of the receptors inhibit cytotoxicity, mostly by turning off NK cells when HLA is exhibited on tumor cells. Current ongoing trials are testing an anti-KIR moAb in conjugation with ipilimumab (NCT01750580) or nivolumab (NCT01714739). Anti-PD1 monoclonal antibodies were also being examined in various novel combinations in a phase I setting, such as nivolumab plus agonistic anti-CD137 moAbs (urelumab, NCT02253992), nivolumab plus anti-LAG-3 (NCT01968109), and cetuximab plus urelumab.

DC Vaccines

DC vaccines have acquired desirable interest due to their ability to induce a robust immunity reaction. As stated before, they are being manufactured via isolation of DCs and loading of tumor antigen *ex vivo*, further followed by reintroduction of DCs into the individuals as a cellular vaccine, usually into the tumor itself or into lymph nodes. In a preclinical study, a DC vaccine was modified using a skin flap transfer treated with sensitized DCs in a rat tumor model. It was observed that the DC-treated group exhibited a reduction in tumor size and an immunological response, defined as elevated levels of IL-2 and IFN-γ (Inoue et al. 2015). A phase I trial was conducted in stage I–IVa patients with HNSCC with no active disease using a DC vaccine loaded with two HLA-A*0201-restricted T-cell-defined p53 peptides alone, plus either a wt p53 helper peptide or nonspecific helper peptide derived from tetanus toxoid. In that study, disease-free survival was 88%, and p53-specific T-cell frequencies were increased in approximately 70% of patients, whereas toxicity was in an acceptable stage (Schuler et al. 2014). Finally, in another study, autologous DCs loaded with apoptotic tumor cells were injected intranodally in patients with advanced HNSCC; immunological responses were satisfactory, and all patients were long-term survivors (Whiteside et al. 2016).

Adoptive T-Cell Therapy (ACT)

As described earlier, ACT is a therapeutic procedure where T cells are isolated from peripheral blood mononuclear cells of individuals or from TILs of primary tumor, and undergo *in vitro* expansion and are re-infused into the individual, with the view to enhance the antitumor immune response. Genetic engineering of T cells before reintroduction, potentially augments function through several autonomous mechanisms. In a phase I study conducted in 17 patients with R/M HNSCC, patients were vaccinated on the thigh with irradiated autologous tumor cells; subsequently, T cells derived from resected inguinal lymph nodes were expanded *in vitro* and reintroduced into the patients. Among the patients enrolled, 6 of 17 patients experienced disease control (To et al. 2000). Significantly, efficacy of ACT is enhanced by cytotoxic chemotherapy. In a retrospective study, ACT was added as experimental therapy in patients with resettable HNSCC receiving induction chemotherapy. Surprisingly, median OS and PFs were improved in patients treated with ACT (Jiang et al. 2015). Finally, ACT has been examined in patients with R/M nasopharyngeal carcinoma. In a phase II study, ACT with EBV-specific CTLs in combination with chemotherapy has shown significant results in demonstrating a 2-year OS of 62.9% (Chia et al. 2014).

Future Directions

HNSCC beholds a heterogeneous group of diseases. To date, standard treatment planning has given mediocre results, and prognosis in individuals with advanced disease is dismal. There is an exuberating content of evidence that the immune system plays a significantly vital role in oncogenesis and tumor evolution; immune-editing is the term used to describe the immune system's defensive role against cancer development. Genetic and epigenetic alterations that are characteristic of all tumors exhibit a controversial set of antigens that the immune system could use to distinguish tumor cells from their normal counterparts. Nevertheless, tumors have the ability to change and modulate the immune system to their favor. Our better understanding of the mechanisms of immune escape has given birth to the development of novel immunotherapies that have shown preliminary promising results in many solid tumors, including HNSCC. A plethora of novel strategies are being explored in clinical trials with the view to improve patient outcome.

REFERENCES

Allen C, Duffy S, Teknos T, et al. Nuclear factor-kappaB-related serum factors as longitudinal biomarkers of response and survival in advanced oropharyngeal carcinoma. *Clin Cancer Res.* 2007;13:3182–3190.

Badoual C, Hans S, Fridman WH, et al. Revisiting the prognostic value of regulatory T cells in patients with cancer. *J Clin Oncol.* 2009;27:e5–e6.

Badoual C, Hans S, Merillon N, et al. PD-1-expressing tumor-infiltrating T cells are a favorable prognostic biomarker in HPV-associated head and neck cancer. *Cancer Res.* 2013;73:128–138.

Badoual C, Hans S, Rodriguez J, et al. Prognostic value of tumor-infiltrating CD4+ T-cell subpopulations in head and neck cancers. *Clin Cancer Res.* 2006;12:465–472.

Basavaraj C, Sierra P, Shivu J, et al. Nimotuzumab with chemoradiation confers a survival advantage in treatment-naive head and neck tumors over expressing EGFR. *Cancer Biol Ther.* 2010;10:673–681.

Baxi S, Fury M, Ganly I, et al. Ten years of progress in head and neck cancers. *J Natl Compr Canc Netw.* 2012;10:806–810.

Bonner JA, Harari PM, Giralt J, et al. Radiotherapy plus cetuximab for squamous-cell carcinoma of the head and neck. *N Engl J Med.* 2006;354:567–578.

Bourguignon LY, Wong G, Earle C, et al. Hyaluronan-CD44v3 interaction with Oct4-Sox2-Nanog promotes miR-302 expression leading to self-renewal, clonal formation, and cisplatin resistance in cancer stem cells from head and neck squamous cell carcinoma. *J Biol Chem.* 2012;287:32800–32824.

Campbell KS, Purdy AK. Structure/function of human killer cell immunoglobulin-like receptors: Lessons from polymorphisms, evolution, crystal structures and mutations. *Immunology.* 2011;132:315–325.

Chen YW, Chen KH, Huang PI, et al. Cucurbitacin I suppressed stem-like property and enhanced radiation-induced apoptosis in head and neck squamous carcinoma–derived CD44(+)ALDH1(+) cells. *Mol Cancer Ther.* 2010;9:2879–2892.

Chia WK, Teo M, Wang WW, et al. Adoptive T-cell transfer and chemotherapy in the first-line treatment of metastatic and/or locally recurrent nasopharyngeal carcinoma. *Mol Ther.* 2014;22:132–139.

Chikamatsu K, Sakakura K, Whiteside TL, et al. Relationships between regulatory T cells and CD8+ effector populations in patients with squamous cell carcinoma of the head and neck. *Head Neck.* 2007;29:120–127.

Cho YA, Yoon HJ, Lee JI, et al. Relationship between the expressions of PD-L1 and tumor-infiltrating lymphocytes in oral squamous cell carcinoma. *Oral Oncol.* 2011;47:1148–1153.

Drake CG, Jaffee E, Pardoll DM. Mechanisms of immune evasion by tumors. *Adv Immunol.* 2006;90:51–81.

Dunn GP, Old LJ, Schreiber RD. The immune biology of cancer immune surveillance and immune editing. *Immunity.* 2004;21(2):137–148.

Ermorken JB, Stohlmacher-Williams J, Davidenko I, et al. Cisplatin and fluorouracil with or without panitumumab in patients with recurrent or metastatic squamous-cell carcinoma of the head and neck (SPECTRUM): An open-label phase 3 randomised trial. *Lancet Oncol.* 2013;14:697–710.

Ferlay J, Shin HR, Bray F, et al. Estimates of worldwide burden of cancer in 2008: GLOBOCAN 2008. *Int J Cancer.* 2010;127:2893–2917.

Ferris RL. Immunology and immunotherapy of head and neck cancer. *J Clin Oncol.* 2015;33:3293–3304.

Fury M, Ou SH, Balmanoukian A, et al. *Clinical Activity and Safety of MEDI4736, an Anti-PD-L1 Antibody, in Head and Neck Cancer.* ESMO Meeting 2014. Poster # 988PD. Abstract ID 5656.

Gajewski TF. The next hurdle in cancer immunotherapy: Overcoming the non-T-cell-inflamed tumor microenvironment. *Semin Oncol.* 2015;42:663–671.

Gildener-Leapman N, Ferris RL, Bauman JE. Promising systemic immunotherapies in head and neck squamous cell carcinoma. *Oral Oncol.* 2013;49:1089–1096.

Hoffmann TK, Dworacki G, Tsukihiro T, et al. Spontaneous apoptosis of circulating T lymphocytes in patients with head and neck cancer and its clinical importance. *Clin Cancer Res.* 2002;8:2553–2562.

Inoue K, Saegusa N, Omiya M, et al. Immunologically augmented skin flap as a novel dendritic cell vaccine against head and neck cancer in a rat model. *Cancer Sci.* 2015;106:143–150.

Jewett A, Tseng HC, Arasteh A, et al. Natural killer cells preferentially target cancer stem cells; role of monocytes in protection against NK cell mediated lysis of cancer stem cells. *Curr Drug Deliv.* 2012;9:5–16.

Jiang P, Zhang Y, Wang H, et al. Adoptive cell transfer after chemotherapy enhances survival in patients with resectable HNSCC. *Int Immunopharmacol.* 2015;28:208–214.

Krishnamurthy S, Nor JE. Head and neck cancer stem cells. *J Dent Res.* 2012;91:334–340.

Lalami Y, Awada A. Innovative perspectives of immunotherapy in head and neck cancer. From relevant scientific rationale to effective clinical practice. *Cancer Treat Rev.* 2016;43:113–123.

Leibowitz MS, Andrade Filho PA, Ferrone S, et al. Deficiency of activated STAT1 in head and neck cancer cells mediates TAP1-dependent escape from cytotoxic T lymphocytes. *Cancer Immunol Immunother.* 2011;60:525–535.

Lukesova E, Boucek J, Rotnaglova E, et al. High level of Tregs is a positive prognostic marker in patients with HPV-positive oral and oropharyngeal squamous cell carcinomas. *Biomed Res Int.* 2014;2014:4–7.

Lyford-Pike S, Peng S, Young GD, et al. Evidence for a role of the PD-1:PD-L1 pathway in immune resistance of HPV-associated head and neck squamous cell carcinoma. *Cancer Res.* 2013;73:1733–1741.

Medzhitov R, Janeway CA Jr. Decoding the patterns of self and nonself by the innate immune system. *Science.* 2002;296:298–300.

Mendes F, Domingues C, Rodrigues-Santos P, et al. The role of immune system exhaustion on cancer cell escape and anti-tumor immune induction after irradiation. *Biochim Biophys Acta.* 2016;1865:168–175.

Näsman A, Romanitan M, Nordfors C, et al. Tumor infiltrating CD8+ and Foxp3+ lymphocytes correlate to clinical outcome and human papillomavirus (HPV) status in tonsillar cancer. *PLoS One.* 2012;7:e38711.

Pardoll DM. The blockade of immune checkpoints in cancer immunotherapy. *Nat Rev Cancer.* 2012;12:252–264.

Qian X, Ma C, Nie X, et al. Biology and immunology of cancer stem(-like) cells in head and neck cancer. *Crit Rev Oncol Hematol.* 2015;95:337–345.

Qian X, Wagner S, Ma C, et al. ALDH1-positive cancer stem-like cells are enriched in nodal metastases of oropharyngeal squamous cell carcinoma independent of HPV status. *Oncol Rep.* 2013;29:1777–1784.

Rampias T, Sasaki C, Weinberger P, et al. E6 and e7 gene silencing and transformed phenotype of human papillomavirus 16-positive oropharyngeal cancer cells. *J Natl Cancer Inst.* 2009;101:412–423.

Raval RR, Sharabi AB, Walker AJ, et al. Tumor immunology and cancer immunotherapy: Summary of the 2013 SITC primer. *J Immunother Cancer.* 2014;2:14.

Reichert TE, Strauss L, Wagner EM, et al. Signaling abnormalities, apoptosis, and reduced proliferation of circulating and tumor-infiltrating lymphocytes in patients with oral carcinoma. *Clin Cancer Res.* 2002;8:3137–3145.

Saloura V, Cohen EE, Licitra L, et al. An open-label single-arm, phase II trial of zalutumumab, a human monoclonal anti-EGFR antibody, in patients with platinum-refractory squamous cell carcinoma of the head and neck. *Cancer Chemother Pharmacol.* 2014;73:1227–1239.

Sato T, Terai M, Tamura Y, et al. Interleukin 10 in the tumor microenvironment: A target for anticancer immunotherapy. *Immunol Res.* 2011;51:170–182.

Schaefer C, Kim GG, Albers A, et al. Characteristics of CD4+CD25+ regulatory T cells in the peripheral circulation of patients with head and neck cancer. *Br J Cancer.* 2005;92:913–920.

Schoenfeld JD. Immunity in head and neck cancer. *Cancer Immunol Res.* 2015;3:12–17.

Schuler PJ, Harasymczuk M, Visus C, et al. Phase I dendritic cell p53 peptide vaccine for head and neck cancer. *Clin Cancer Res.* 2014;20:2433–2444.

Seidel UJ, Schlegel P, Lang P. Natural killer cell mediated antibody-dependent cellular cytotoxicity in tumor immunotherapy with therapeutic antibodies. *Front Immunol.* 2013;4:76.

Seiwert TY, Gupta S, Mehra R, et al. Antitumor activity and safety of pembrolizumab in patients (pts) with advanced squamous cell carcinoma of the head and neck (SCCHN): Preliminary results from KEYNOTE-012 expansion cohort. *J Clin Oncol.* 2015;33:6.

Siddiqui SA, Frigola X, Bonne-Annee S, et al. Tumor-infiltrating Foxp3-CD4+CD25+ T cells predict poor survival in renal cell carcinoma. *Clin Cancer Res.* 2007;13:2075–2081.

Spranger S, Gajewski T. Rational combinations of immunotherapeutics that target discrete pathways. *J Immunother Cancer.* 2013;1:16.

Sun W, Li WJ, Wu CY, et al. CD45RA-Foxp3 high but not CD45RA+Foxp3 low suppressive T regulatory cells increased in the peripheral circulation of patients with head and neck squamous cell carcinoma and correlated with tumor progression. *J Exp Clin Cancer Res.* 2014;33:35.

Tang AL, Owen JH, Hauff SJ, et al. Head and neck cancer stem cells: The effect of HPV—An in vitro and mouse study. *Otolaryngol Head Neck Surg.* 2013;149:252–260.

To WC, Wood BG, Krauss JC, et al. Systemic adoptive T-cell immunotherapy in recurrent and metastatic carcinoma of the head and neck: A phase 1 study. *Arch Otolaryngol Head Neck Surg.* 2000;126:1225–1231.

Tsushima F, Tanaka K, Otsuki N, et al. Predominant expression of B7-H1 and its immunoregulatory roles in oral squamous cell carcinoma. *Oral Oncol.* 2006;42:268–274.

Van der Burg SH. Immunotherapy of human papilloma virus induced disease. *Open Virol J.* 2012;6:257–263.

Varilla V, Atienza J, Dasanu CA. Immune alterations and immunotherapy prospects in head and neck cancer. *Expert Opin Biol Ther.* 2013;13:1241–1256.

Vasilakopoulou M, Rampias T, Sasaki C, et al. Effect of PDL-1 expression on prognosis in head and neck squamous cell carcinoma. *J Clin Oncol.* 2013;31:abstr 6012.

Vermorken JB, Mesia R, Rivera F, et al. Platinum-based chemotherapy plus cetuximab in head and neck cancer. *N Engl J Med.* 2008;359:1116–1127.

Vesely MD, Kershaw MH, Schreiber RD, et al. Natural innate and adaptive immunity to cancer. *Annu Rev Immunol.* 2011;29:235–271.

Wang T, Niu G, Kortylewski M, et al. Regulation of the innate and adaptive immune responses by Stat-3 signaling in tumor cells. *Nat Med.* 2004;10:48–54.

Weinberger PM, Yu Z, Haffty BG, et al. Molecular classification identifies a subset of human papillomavirus–associated oropharyngeal cancers with favorable prognosis. *J Clin Oncol.* 2006;24:736–747.

Whiteside TL. Down-regulation of zeta-chain expression in T cells: A biomarker of prognosis in cancer? *Cancer Immunol Immunother.* 2004;53:865–878.

Whiteside TL, Ferris RL, Szczepanski M, et al. Dendritic cell-based autologous tumor vaccines for head and neck squamous cell carcinoma. *Head Neck.* 2016;38 Suppl 1: E494–E501.

Woo SR, Corrales L, Gajewski TF. The STING pathway and the T cell-inflamed tumor microenvironment. *Trends Immunol.* 2015;36:250–256.

Yu H, Pardoll D, Jove R. STATs in cancer inflammation and immunity: A leading role for STAT3. *Nat Rev Cancer.* 2009;9:798–809.

Zhang M, Kumar B, Piao L, et al. Elevated intrinsic cancer stem cell population in human papillomavirus-associated head and neck squamous cell carcinoma. *Cancer.* 2014;120:992–1001.

Zou W, Chen L. Inhibitory B7-family molecules in the tumour microenvironment. *Nat Rev Immunol.* 2008;8:467–477.

7 Check the Cancer Before It Checks You Out

Bhargavi Dasari and Izaz Shaik

CONTENTS

INTRODUCTION

Interest in cancer has grown during the past century as infectious diseases have increasingly been controlled as the result of improved sanitation, vaccinations, and antibiotics. Although the interest is recent, cancer is not a new disease. It has existed for many centuries. According to worldwide cancer mortality statistics in 2012, it was estimated that there were 8.2 million deaths reported due to cancer. The world age standardized mortality rate shows that there are about 126 cancer deaths for every 100,000 men in the world, and 83 for every 100,000 females. In 2012, there were an estimated 14.1 million new cases of cancer in the world: 7.4 million (53%) in males and 6.7 million (47%) in females, giving a male:female ratio of 10:9. Survival rates are approximately 50%. However, early detection followed by appropriate treatment can increase cure rates to about 80%, and greatly improve quality of life by minimizing extensive, debilitating treatments. Hence, screening and early diagnosis play an important role in the prevention and treatment of cancer.

This chapter describes the importance of early detection of cancer and screening methods of cancer both by self as well by the clinician. Also, various noninvasive diagnostic methods are described.

TOLUDINE BLUE

Toludine blue, discovered during the 1960s, is a basic metachromatic dye of the thiazine group that shows affinity for the perinuclear cisternae of DNA and RNA (Herlin et al. 1983). Allen et al. (1949) introduced the therapeutic use of tolonium chloride as an IV antiheparin agent. In 1960 Sherwin suggested (Strong et al. 1968) the use of tolonium chloride for *in vivo* staining of suspicious lesions of the oral cavity that might stain tumor cells and normal mucosa or leukoplakia differentially (Kozlovsky et al. 1994). Reichart (1963) proposed the use of toluidine blue as a vital stain to disclose dysplasia and carcinoma in situ of the uterine cervix. Reichart proposed that the affinity and intensity of toluidine blue vital staining were dependent on the number of nuclei per unit area of tissue, since toluidine blue primarily stains nucleic acids (Upadhyay et al. 2011). This *in vivo* demonstration of oral cavity lesions is based on the observation that the topical application of toluidine blue, an acidophilic metachromatic nuclear stain, will stain an area of carcinoma in situ or invasive carcinoma, whereas normal mucosa will not stain (Myers 1970). The stain is incorporated nonspecifically into cells and tissues following 1% aqueous application and differentially stains only those tissues rich in nucleic acids after 1% acetic acid decolorization. Malignant, premalignant, ulcerated, and inflamed tissues are rich in nucleic acids and therefore are expected to stain positively (Upadhyay et al. 2011). Niebel and Chomet (1964) were the first to report vital application of toluidine blue for detection of premalignant and malignant lesions of the oral cavity. These investigators confirmed the property of toluidine blue/decolorization to verify clinically suspicious lesions as neoplastic, to delineate margins of premalignant and malignant growth, and to detect unnoticed or satellite tumors. False-positive results were infrequent, and false-negative results were not reported by these investigators. Vahidy et al. (1972) reported a sensitivity of 86% and a specificity of 76% after excluding numerous doubtful lesions where staining properties could not be readily judged as either positive or negative. The positive attributes of this technique were considered by most to be greater than the problem of the high incidence of false positives (Upadhyay et al. 2011).

MECHANISM

Tolonium chloride is a metachromatic dye that stains mitochondrial DNA, cells with greater-than-normal DNA content or altered DNA in dysplastic and malignant cells.

Its use *in vivo* is based on the fact that dyslastic and anaplastic cells contain quantitatively more nucleic acids and increased mitoses than normal surrounding epithelium (Vercellino et al. 1985). Another mechanism appears to be greater penetration and temporary retention of the dye in the intercellular spaces of rapidly dividing cells *in vivo* RNA (Herlin et al. 1983; Kozlovsky et al. 1994).

However, the mechanism by which the dye differentially stains malignant and dysplastic tissues remains unclear. Epstein et al. (1997) showed that the use of toluidine blue is more sensitive than clinical examination alone, and compared to iodine staining (sensitivity of 73%), toluidine blue (sensitivity 91.2%) has yielded better results (Kozlovsky et al. 1994).

The formula of toluidine blue solution is as follows:

- Toluidine blue 1 g
- Acetic acid 10 cc
- Absolute alcohol 4.19 cc
- Distilled water 86 cc
- pH adjusted to 4.5 (Epstein et al. 1992)

PROCEDURE

Patients rinsed the oral cavity with water for 20 seconds to remove debris prior to rinsing with 1% acetic acid for 20 seconds. Toluidine blue (1% W/W) was applied as an oral rinse for 20 seconds, and then 1% acetic acid was used for 20 seconds to eliminate mechanically retained stain (Figure 7.1). Lesions that showed dark blue staining were considered to be positive for premalignant or malignant tissue, while those with light staining, or totally not colored, were considered negative (Allegra et al. 2009).

The use of toluidine blue as a screening test is a valuable adjunct to inspection and palpation of the oral cavity in the search for malignant lesions. It is of great use in follow-up after radiation therapy, or after other treatment for mucous membrane lesions since unnecessary biopsies for postradiation or traumatic ulcerations can be avoided. The lack of false positives or false negatives enhances its value as a screening test. The test itself is remarkably simple and requires only cleansing of the ulcer so that the stain will come in contact with the surface of the ulcer. Earlier studies have shown that normal mucous membrane does not stain; therefore, one must not depend on this test for those tumors that spread without involvement of overlying mucous membrane. The test is to be used as an adjunct and as a screening test, and is not a substitute for biopsy and close observation, which are the best ways to deal with lesions of the oral cavity (Myers 1970).

Giler et al. have used toluidine blue in the diagnosis of malignant gastric lesions (Epstein et al. 1992).

Lundgren et al. studied a series of 272 glottic lesions with toluidine blue application that were followed by biopsy. The sensitivity in detection of malignant lesions

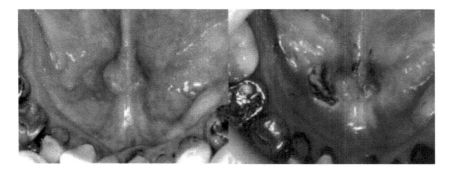

FIGURE 7.1   Before and after images of use of 1% acetic acid for 20 seconds.

was 91%, and the specificity was 51%. Inflamed and ulcerated areas retained dye (Epstein et al. 1992).

In the false-positive reactions, Strong (1968) first suggested *in vivo* staining of possible malignant lesions of the oral cavity. Several studies confirmed that nondys-plastic epithelium failed to retain toluidine blue, but that malignant lesions stained intensely (Epstein et al. 1992). Patton et al. (2008) conducted a study on adjuvant techniques for early detection of oral premalignant and malignant lesions for detection of malignancy, and concluded that TB is effective as a diagnostic adjunct for use in high-risk populations and suspicious mucosal lesions.

Güneri et al. (2010) conducted a study to investigate the utility of toluidine blue and brush cytology in patients with clinically detected oral mucosal lesions. In conclusion, adjunct diagnostic methods decreased the level of uncertainty for the diagnosis of oral malignancies and lichenoid dysplasias when applied as adjuncts to clinical examination (Güneri et al. 2011).

Upadhyay et al. (2011) conducted a study on the use of toluidine blue for the detection of potentially malignant oral lesions (PMOLs) and malignant lesions. They concluded that toludine blue staining should not blindly direct the clinician's opinion, strongly discouraged the use of toluidine blue as a screening test, and noted that the results should be interpreted with caution (Kozlovsky et al. 1994).

Rahman et al. (2012) concluded that 1% toluidine blue and cytology have high sensitivity, specificity, and accuracy in detecting oral premalignant lesions and oral squamous cell carcinoma and can be used as an adjunct in the early detection of oral lesions.

LUGOL'S IODINE

The vital dyes are an auxiliary technique used *in vivo* in order to evidence suspicious lesions and/or to better define the lesion margins and extension. These stains are capable of penetrating living cells and binding to specific biological structures. Lugol's iodine, also known as Lugol's solution, first made in 1829, is a solution of elemental iodine and potassium iodide in water, named after the French physician Lugol (1786–1851) (Petruzzi et al. 2010).

Lugol's iodine solution is often used as an antiseptic and disinfectant, for emergency disinfection of drinking water, and as a reagent for starch detection in routine laboratory and medical tests. Original Lugol's solution consists of

- 5 g iodine (I2) and
- 10 g potassium iodide (KI) mixed with
- 85 mL distilled water, to make a brown solution with a total iodine content of 150 mg/mL

Other names for Lugol's solution are

- I_2KI (iodine–potassium iodide)
- Markodine
- Strong solution (systemic)
- Aqueous iodine solution BCP (Petruzzi et al. 2010)

Principle

The principle of iodine staining is that iodine reacts with glycogen in the cytoplasm and the reaction, known as the iodine–starch reaction, is visualized by a color change. Tissue glycogen content is related to the degree of keratinization; glycogen content being inversely proportional to the degree of keratinization, because glycogen plays a key role in keratinization. Moreover, the loss of cellular differentiation and the enhanced glycolysis in cancer cells do not promote the iodine–starch reaction. During mucosal examination, Lugol's iodine is applied on the suspicious lesions: normal mucosa stains brown or mahogany due to its high glycogen content, while dysplastic tissue does not stain, and appears pale compared to the surrounding tissue. The vital dye with Lugol's solution is also called Schiller's test thanks to the Austrian–American gynecologist and pathologist who first described the technique in 1933 in gynecological lesions. In the 1960s, this vital staining was first used to investigate esophageal diseases. To date, Schiller's test is used as a diagnostic aid in the detection of esophageal, gastrointestinal, and gynecological suspicious lesions associated with endoscopic and colposcopic techniques. In the oral cavity, efficacy of Lugol's iodine staining is restricted to non-keratinized mucosa (buccal, vestibule, ventral surface and margins of the tongue, oral floor), while for the other mucosal areas other techniques should be used for the detection of early carcinoma and its margins (Petruzzi et al. 2010).

The formula for Lugol's iodine solution is as follows:

- Iodine 2 g
- Potassium iodide 4 g
- Distilled water 100 cc (Epstein et al. 1992)

Procedure

The lesion areas were applied prior with 1% acetic acid with a cotton bud for 20 seconds and further rinsed with water. Lugol's iodine was applied with a cotton bud for 10–20 seconds (Nagaraju et al. 2010) (Figure 7.2).

Silverman et al. (1971) did not find the glycogen content of epithelial cells to be related to inflammation. However, previous studies conducted by Doyle et al. in 1968 had shown a correlation between the degree of inflammation and glycogen content.

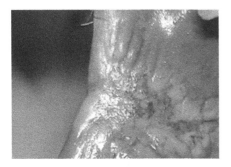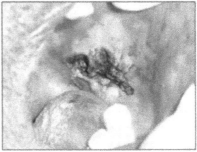

FIGURE 7.2 Diagnosis of leukoplakia after application of Lugol's iodine.

This may limit the use of Lugol's iodine in keratinized lesions, and Lugol's iodine uptake should be reviewed with caution (Epstein et al. 1992).

Schiller (1933) was the first to describe the use of Lugol's solution to detect abnormal cervical epithelium (Epstein et al. 1992).

Many studies have shown that glycolysis is elevated in cancer cells, and such cellular areas with elevated glycolysis are generally seen as unstained lesions (USLs) during vital iodine staining. Since vital iodine staining was first performed on the cervix and esophageal mucosa, it has been combined routinely with endoscopy of the upper gastrointestinal tract, and it has contributed markedly to the early detection of esophageal cancer. Compared with the upper gastrointestinal tract, the thickness of the mucosal epithelium varies in the oral cavity (100–500 mm), and the presence of teeth, saliva, and movement of the tongue make it difficult to observe tissue samples under similar conditions (Maeda et al. 2010).

Maeda et al. (2010) compared three different concentrations of Lugol's iodine (3%, 5%, and 10% iodine glycerine) in the detection of clear boundary lesions and in the distinction of mild, moderate, and severe dysplasia using a colorimetric analysis of the USLs. They concluded that the most effective method was to stain the lesion with 3% Lugol's solution followed by 5% Lugol's solution. In their successive study, they realized a color chart, prepared on the basis of lightness and hue values (Maeda et al. 2009, 2010) and macroscopically obtained by tongue staining. McMahon et al. (2010) evaluated, in their comparative study, the efficacy of Lugol's solution in the determination of extension of suspicious lesions. These authors demonstrated significantly less incidence of dysplasia or carcinoma among the margins of resection of lesion prestained with Lugol's solution.

OPTICAL SYSTEMS

The interaction of light with tissues may highlight changes in tissue structure and metabolism. Optical spectroscopy systems to detect changes relying on the fact that the optical spectrum derived from a tissue will contain information about the histological and biochemical characteristics of that tissue. Such optical adjuncts may assist in the identification of mucosal lesions, including premalignant lesions and oral squamous cell carcinomas; may assist in biopsy site selection and enhance visibility of the surface texture and margins of lesions; and may also assist in the identification of cellular and molecular abnormalities not visible to the naked eye on routine examination. There are a number of optical systems that can yield similar types of information approaching the detail of histopathology, and theoretically at least, in a more quantifiable and objective fashion, in real-time, noninvasively, and in situ (Scully et al. 2008).

VIZILITE/CHEMILUMINESCENCE

ViziLite is one commercially available tool that makes use of acetic-acid-induced whitening of oral tissues and chemiluminescent light to improve identification, evaluation, and monitoring of white oral mucosal abnormalities in populations at increased risk for oral cancer.

The ViziLite tool, using chemiluminescent light, is marketed to be used as an adjunct to traditional oral examination by incandescent light (Farah and

McCullough 2007). The term *chemiluminescence* refers to the emission of light from a chemical reaction. Chemiluminescent reactions emit light of varying degrees of intensity (Saravanan and Siar 2003). The manufacturer claims that following the application of a cytoplasmic dehydration agent such as an acetic acid solution, leukoplakic lesions are better visualized due to changes in their refractile properties. This occurs in atypical nonkeratinized squamous epithelium due to an increase in the nuclear-to-cytoplasmic ratio of the cells. According to the manufacturer, as ViziLite passes over oral tissue that has been treated with the acetic acid rinse solution, normal healthy tissue will absorb the light and appear dark with a blue hue, while abnormal tissue will appear white (aceto-white lesion). The chemiluminescent system used in ViziLite has not as yet been identified, although it is most likely based on the peroxy-oxalate system, with the outer flexible capsule containing acetyl salicylic acid and the inner fragile glass vial containing hydrogen peroxide. These chemicals react to produce a light of blue–white color with a wavelength between 430 and 580 nm (Saravanan and Siar 2003) (Figure 7.3).

The kit consists of a 1% acetic acid solution, a capsule (which emits light), a retractor, and manufacturer's instructions. The capsule is formed by an outer shell of flexible plastic and an inner vial of fragile glass. Although the company has not provided data on its precise composition, some authors reported that the outer capsule may contain acetylsalicylic acid and the inner vial hydrogen per-oxide. For its activation, the capsule is bent, breaking the glass vial so that the chemical products react and produce a bluish-white light with a wavelength of 430–580 nm that lasts for around 10 minutes. The patient performs a 1-minute mouthwash with the acetic acid solution to remove the glycoprotein barrier and slightly dry the mucosa. The intensity of ambient light is then dimmed, and a diffuse bluish-white chemiluminescent light is applied. Normal cells absorb the light and have a bluish color, whereas the light is reflected by abnormal cells with a higher nucleus:cytoplasm ratio and by epithelium with excessive keratini-zation, hyperparakeratinization, and/or significant inflammatory infiltrate, which appear aceto-white with brighter, more marked, and more distinguishable borders (Trullenque-Eriksson et al. 2009).

Early detection of mucosal lesions can be enhanced by the use of a dilute acetic acid rinse and observation under a chemiluminescent light (ViziLite). In one study,

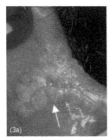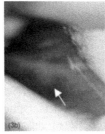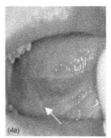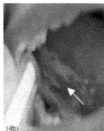

FIGURE 7.3   Lesion diagnosis with ViziLite.

78 of 100 patients who presented for dental screening were examined by COE (under incandescent light). After a 1-minute rinse with 1% acetic acid and then using the ViziLite, 57 clinically diagnosable benign lesions (e.g., linea alba, leukoedema) and 29 clinically undiagnosable lesions were found initially. After the acetic acid rinse again, six additional diagnosable lesions (linea alba) and three undiagnosable lesions were found. In a multicenter study, increased visibility of lesions visible by COE was reported. In that and other studies, ViziLite revealed occasional lesions not seen under COE, but occasionally, the converse has been the case, so the jury again is out on the real benefits (Scully et al. 2008).

The study by Ram and Siar (2005) assessed the value of a commercially available chemiluminescent light kit or ViziLite over 1% tolonium chloride as a diagnostic aid in the early detection of oral cancer and PMELs and concluded that chemiluminescent light or ViziLite is useful as an adjunctive diagnostic tool for the detection of oral cancer and PMELs and follow-up of subjects treated for the same. Saravanan and Siar (2003) also stated that ViziLite is more efficacious than tolonium chloride for early detection of oral cancer and premalignant lesions. However, Oh and Laskin (2007) concluded a study to evaluate early detection by using dilute acetic acid rinse and observation under a chemiluminescent light and concluded that acid rinse accentuated some lesions but the overall detection rate was not significantly improved. The chemiluminescent light produced reflections that made visualization more difficult and thus was not beneficial (Oh and Laskin 2007). Farah and McCullough (2007) considered that ViziLite cannot discriminate between malignant, benign, and inflammatory oral lesions.

In fact, the main drawback of this technique is its low specificity and the high rate of false positives, which could give rise to unnecessary biopsies. Its combination with toluidine blue has been proposed (ViziLite Plus) in order to reduce the number of false positives without increasing the rate of false negatives (Trullenque-Eriksson et al. 2009).

Other limitations are its high cost and its inability to indicate the appropriate site for a biopsy. It has also been pointed out that there is no clinical evidence to justify the additional cost of the system and that detection by an expert clinician remains essential (Trullenque-Eriksson et al. 2009).

AUTOFLUORESCENCE IMAGING

Autofluorescence imaging has been used successfully to rapidly and noninvasively distinguish malignant oral lesions from surrounding tissue in the oral cavity. A low-cost device for visualization of oral autofluorescence was used to identify high-risk precancerous and cancerous lesions with 98% sensitivity and 100% specificity based on the loss of fluorescence in abnormal sites compared with normal tissue. This device is now commercially available as VELSCOPE (Pavlova et al. 2008). Early detection of mucosal lesions can be enhanced by the use of fluorescence. All tissues have a tendency to glow (fluoresce) in the dark, either spontaneously (autofluorescence) or if an external sensitizer is applied to the tissues. The tissue fluoresces due to the presence of fluorescent chromophores (fluorophores) within the cells. Commonly detected fluorophores include nicotine adenine dinucleotide hydrogenase, collagen, elastin, flavin adenine dinucleotides, hemoglobin and vascular supply, and oral microbial flora, and they vary in different tissues including in different sites in the mouth. Tissue changes can affect the fluorophores and tissue fluorescence, and this may facilitate detection of lesions not detectable with the naked eye under normal incandescent white light. Fluorescence and changes suggestive of PML or OSCC can already be detected using commercially available photographic techniques (e.g., Storz, Pentax, Zillix), but most of these also have relatively low sensitivity and specificity. Preliminary studies using direct visualization (VelScope), however, have been very encouraging when assessed in patients with OSCC (Scully et al. 2008).

Early work by Onizawa et al. (1996) on analyzing autofluorescence in oral mucosa detected fluorescence from malignant lesions, between the orange and red parts of the spectrum; this study suggested that the concentration of blood constituents along with protoporphyrin are responsible for this autofluorescence, which seemed to increase during tumor progression and decrease during tumor regression. Subepithelial stromal collagen fibers are the predominant source of autofluorescence in oral mucosa. These are confirmed in some fluorescence studies, which reported decreased green fluorescence, attributed to collagen degradation, and increased red fluorescence, attributed to release of porphyrin. This change in fluorescence signal between normal and neoplastic lesion was confirmed in oral cavity lesions, and the possibility of delineating the tumor margin based on this fluorescence change was studied by Poh et al. (2006). They illuminated the oral cavity with blue light from a hand-held device, which was later named VELscope, and visualized autofluorescence. Loss of green fluorescence in cancerous tissues was noted, and physicians manually demarcated the tumor margins based on this green autofluorescence change, and the microdissected tumor margin biopsies underwent loss of heterozygosity (LOH) analysis to confirm the demarcation (Figure 7.4). This shows that simple, noninvasive autofluorescence detection can effectively detect even occult neoplastic lesions, which might otherwise go unnoticed using white light visualization (Olivo et al. 2011).

(a)   (b)

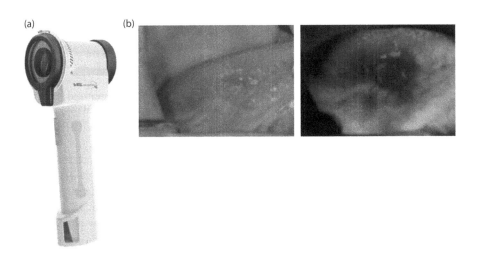

FIGURE 7.4   (a) VELscope, (b) Visualization of precancerous lesion using VELscope.

Another important study by Richards-Kortum et al. (2004) quantified the green–red autofluorescence using digital image processing and delineated areas of oral dysplasia and carcinoma. This study reported that autofluorescence obtained by illumination of tissues with light of 405 nm wavelength produced the highest discrimination. They calculated the ratio of red-to-green intensity for each pixel digitally within the autofluorescence image, and classified neoplastic with nonneoplastic tissues based on a set threshold value. Disease probability maps, which represent areas where probability of tissues being neoplastic is high, were generated based on this ratio. These regions were confirmed with histological analysis, and validation of this technique resulted in 100% sensitivity and 91.4% specificity. The experimental setups reported for autofluorescence imaging of oral cavity are simple and similar. Autofluorescence images are obtained by exciting tissues generally with light in the 375–440 nm range, but the aforementioned study by Richards-Kortum et al. reported that excitation at 405 nm best discriminates normal from cancerous lesions. A wide-field optical microscope is used to collect digital autofluorescence images with a color CCD camera. Patients are usually imaged in outpatient settings or in the operating room under mild anesthesia (Olivo et al. 2011).

The parameter most of the autofluorescence imaging studies used is the weakened green fluorescence, suspected from connective tissue degradation, and increased red fluorescence, suspected from accumulated porphyrin. Though studies by Betz et al. (1999) reported that red fluorescence is not really specific for malignancies, the ratio of red-to-green autofluorescence was still found to be a good classifier (Olivo et al. 2011).

Though many studies reported autofluorescence imaging to be an objective technique for the detection and demarcation of neoplastic oral cavity lesions, finding its capability of detecting invisible tumors within high-risk populations, such as heavy tobacco users, patients with history of oral cavity cancer, etc., should be the next step for consideration as a robust screening technique. The VELscope has provided the precedent for clinically viable autofluorescence viewing of oral lesions, but a robust system to quantify the autofluorescence and be able to stage tumors should be the way forward. Though studies by Richards-Kortum et al. did quantify autofluorescence through a ratio of green-to-red fluorescence, more extensive studies with wide disease prevalence must be undertaken in order to classify benign and malignant lesions, and to be considered for screening a large population (Olivo et al. 2011).

Evaluation of autofluorescence imaging with VELscope can assist in the identification of malignant and potentially malignant oral lesions from normal mucosa in high-risk patients but does not help in discriminating benign lesions from malignant or premalignant mucosal conditions. Huber (2008) also showed that VELscope interpretation did not enhance or alter clinical management of suspicious lesions. Huber's study reported that several commonly occurring conditions, such as mucosal pigmentations, ulcerations, irritations, and gingivitis, were associated with a loss of fluorescence using VELscope. However, one of the major disadvantages of autofluorescence imaging is still low specificity in the detection of premalignant lesions and early stage cancer. This disadvantage could be overcome with the appearance of new and improved technologies in autofluorescence, such as the addition of backscattered light analysis, ultraviolet spectra, fluorescence-reflectance, or dual digital systems (Olivo et al. 2011).

Only when screened early for oral cancer risks is intervention possible and the treatment more effective. In this regard, autofluorescence imaging comes close to being used as a robust oral cavity cancer screening and detection technique. Early detection of small cancer foci is required to improve the survival rate. Here, autofluorescence might become useful as *in vivo* biochemical changes of cancer metabolism can be shown by fluorescence. However, several nondysplastic lesions may be nonfluorescent, and occasional false-positive results have been reported. Therefore, fluorescence agents are being explored to further improve oral cancer diagnostics (Olivo et al. 2011).

FLUORESCENCE DIAGNOSIS/LIGHT-INDUCED FLUORESCENCE SPECTROSCOPY

Fluorescence diagnosis (FD) is of increasing interest in oral cancer detection. Conventional white light endoscopy that is currently used for oral cancer detection may fail to detect small and flat mucosal neoplasms, and thus, they are frequently overlooked during routine examination. Accurate detection and demarcation of early neoplasms followed by efficient treatment can significantly improve the survival rates of oral cancer patients. Though previous studies have evaluated the use of vital staining with Lugol's iodine and toluidine blue for improving the detection of early oral neoplasms, these methods are not yet clinically useful due to high false-positive or false-negative results. FD using a fluorescence endoscopy system is a technique to visualize the neoplastic lesions in a tumorous organ after topical or systemic application of a tumor-selective photosensitizer. Exact demarcation of tumor margins using this technique could contribute to optimum results in surgical excision and reconstruction. Numerous photosensitizers are being investigated as fluorescent markers for *in vivo* detection and demarcation of tumors. One of the most promising photosensitizers for oral cancer diagnosis is 5-ALA. 5-ALA is a precursor in the heme biosynthetic pathway of nucleated cells. It is metabolized by certain endogenous enzymes to produced PPIX, which is an endogenous photosensitizer. PPIX is an intermediate by-product in the heme biosynthetic pathway, and it preferentially accumulates in tumor cells due to changes in the activity of two main enzymes, porphobilinogen deaminase and ferrochelatase. 5-ALA–induced PPIX fluorescence in the tissue is limited to the epithelium, and both normal and dysplastic epithelium has shown PPIX fluorescence suggesting the usefulness of PPIX fluorescence in the determination of superficial tumor margins (Olivo et al. 2011).

Images of white light (left) and ALA-induced PPIX fluorescence (right) of an oral lesion acquired using a fluorescence endoscopy system (Figure 7.5).

Another photosensitizer, hypericin, has shown excellent fluorescence diagnostic properties in oral cavity cancers. Hypericin is a plant-based photosensitizer that accumulates in abnormal cells including tumor cells. Hypericin fluorescence diagnostic imaging as a technique can facilitate guided biopsies in the clinic, thereby reducing the number of biopsies taken. It can also provide visualization of tumor margins during surgical procedures and assist for same-day diagnosis in the clinic. Studies have reported that hypericin fluorescence can provide improved specificity and is subject to reduced photobleaching compared to 5-ALA (Olivo et al. 2011).

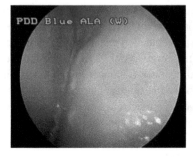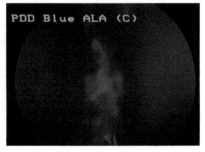

FIGURE 7.5 Fluorescence endoscopy—Visualization of margins of the lesion by fluorescence.

Images of white light (left) and hypericin image (right) of an oral lesion acquired using a fluorescence endoscopy system (Figure 7.6).

Fluorescence photography detected OSCC with a sensitivity of 91% and specificity of 85%. The relatively low sensitivity and specificity of autofluorescence can be markedly improved by adding an exogenous chemical such as aminolevulinic acid (ALA), plied systemically or topically. Typically in the oral cavity, a mouthwash is applied, and the ALA is taken up into the cells and metabolized to protoporphyrin, which fluoresces. Interrogation with blue light results in a fluorescence signal that is then captured using a CCD camera that is mosaic gated and allows specific measurement of red and green fluorescence (Scully et al. 2008).

Laser-induced fluorescence (LIF) spectroscopy has been developed for the diagnosis of cancer using an algorithm based on a nonlinear maximum representation and discrimination feature (MRDF) method and used to examine OSCC in the hamster buccal pouch model shows increased fluorescence in malignant areas (Scully et al. 2008).

RAMAN SPECTROSCOPY

Raman spectroscopy has been used in biology and biochemistry to study the structure and dynamic function of important molecules. The molecular structures of proteins and lipids differ among neoplastic and normal tissues, and therefore, Raman spectroscopy has been considered as a promising tool for diagnosing cancer, in real time and in a less

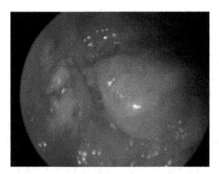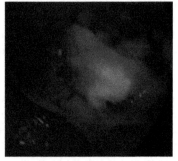

FIGURE 7.6 Hypericin—Diagnosis of oral cavity.

invasive way. The major limitation of Raman spectroscopy in the biological sample is the strong fluorescence emission, which obscures the weak Raman signal. Raman spectroscopy looks at the vibrational changes in tissue that parallel changes in chemical composition, and it is sensitive (for example) to changes in DNA content. Raman spectroscopy is widely used in chemical analysis and is based on "inelastic" light scattering since the detected wavelengths are different from those of the applied light. Fourier transform infrared (FTIR)/Raman spectroscopy has been successfully applied for the diagnosis of OSCC in the hamster cheek pouch model with 100% sensitivity and 55% specificity (Scully et al. 2008). Manoharan et al. (1992) reported that Raman spectroscopy has a potential to identify biological markers associated with the malignant changes, which may provide useful qualitative and quantitative information for use in tumor classification, grading, and evolution, helping in diagnosis and prognosis of the cancer. Choo et al. (1995) reported that many of the biochemical changes in a neoplastic tissue, such as increases in protein content, increases of nucleic acid, and decreases in lipid components, could be detected by Raman spectroscopy (Oliveira et al. 2004).

ORTHOGONAL POLARIZATION SPECTRAL (OPS) IMAGING

Orthogonal polarization spectral (OPS) imaging for *in vivo* visualization of human microcirculation facilitates high-resolution images of the oral mucosa (Scully et al. 2008) (Figure 7.7). The microcirculation plays a crucial role in the interaction between blood and tissue in both the physiological and pathophysiological states. Analysis of microvascular blood flow alterations provides a unique perspective to study processes at the microscopic level in clinical medicine (Fagrell and Intaglieta 1997). Despite the critical role of microcirculation in numerous diseases including diabetes (Tooke 1996), hypertension, sepsis (Lehr et al. 2000), and multiple organ failure, methods for direct visualization and quantitative assessment of the human microcirculation at the bedside are limited (Černý et al. 2007).

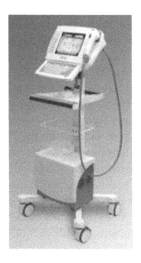

FIGURE 7.7   OPS imaging machine.

MECHANISM

Polarized light (550 ± 70 nm) traveling through a light guide and passing through a series of lenses is absorbed by hemoglobin (Hb) in erythrocytes, thus projecting an image of circulating dark bodies flowing through tissue vasculature. The light wavelength of 550 ± 70 nm is the isobestic point for Hb. Through the use of this spectroscopic method, OPS imaging technology produces high-contrast images of the microcirculation. OPS imaging technology is incorporated into a small hand-held device. This technology works by illuminating the subject or target tissue with light that has been linearly polarized in one plane, while imaging the remitted light through a second polarizer (analyzer) oriented in a plane precisely orthogonal to that of illumination. OPS imaging consists of a probe with a small lens attached to a halo-gen light source with a polarizing filter selecting light at a wavelength of 550 ± 70 nm and a dichroic mirror. The light moves from the lens to the tissue of interest where it is partly reflected back by the tissue and partly scattered inside the tissue. The scattered light, having lost its polarization, will be able to pass through a polarizing filter that is positioned orthogonal to the first polarizing filter behind the dichroic mirror. After the received light passes these specified lenses and polarized filters, it is caught by a CCD camera and recorded as an image. Because parts of the incident light are absorbed by Hb, the red blood cells can be seen as dark gray structures on a brighter background on a high-resolution video monitor. The signals are digitally recorded, which makes it possible to analyze them offline. All measurements were done using a 5× magnifying lens, resulting in a 325× magnification on the computer monitor using a 640×480 screen resolution. To provide a sterile barrier between the patient and device and to ensure a fixed focus distance, a sterile plastic disposable cap was placed around the probe and its lens. The probe was carefully put on the area of interest without applying pressure. All measurements were recorded on a digital video recorder and viewed on a video monitor (Lindeboom et al. 2006).

OSCC are characterized by chaotic and dilated vessels accompanied by numerous areas of hemorrhage, and this may be detectable by OPS (Scully et al. 2008). The morphology of microvessels in the oral squamous cell carcinoma tumors were characterized by irregular, tortuous (megacapillaries), and abnormal vessels. Moreover, the prominent extravasation of red blood cells was a good indicator of intratumor hemorrhaging. This process of capillary hemorrhaging and/or thromboses in disease can be considered as the end stage of progressive capillary enlargement (Lindeboom et al. 2006).

Many studies examining different areas of the oral cavity focus on the size of the tumor and do not take into account the possible associations of and differences in the location or orientation of the tumor and vascularity. The angiogenic requirements of tumors arising in richly vascularized sites like the tongue or buccal mucosa might be less than those surrounding the tumor in less well-vascularized sites. Abnormal microvascular observations in oral squamous cell carcinoma can be considered significant because the presence of these microvascular abnormalities is regarded as an important factor influencing tumor growth and metastasis. The amount of information as such may be a useful prognostic indicator in oral squamous cell carcinoma and may help determine strategies by identifying a subgroup of patients who could benefit from prophylactic neck surgery or radiotherapy regardless of clinical node status. Controversy remains as to whether a correlation exists between tumor vascularity and metastasis or recurrence (Lindeboom et al. 2006).

OPTICAL COHERENCE TOMOGRAPHY

Optical coherence tomography (OCT), first applied in 1991 by Huang et al., is a noninvasive, interferometric (superimposing or interfering waves) tomographic imaging modality that allows millimeter penetration with micrometer-scale axial and lateral resolution. The time-resolved technique is extensively used clinically in ophthalmology (Figure 7.8). OCT has been applied in the head and neck in an attempt to detect areas of inflammation, dysplasia, and cancer; results were promising, but some studies suffered from poor-resolution images and poor penetration depth (Jerjes et al. 2008).

The first application of low-coherence interferometry in the biomedical optics field was for the measurement of eye length. Adding lateral scanning to a low-coherence interferometer allows depth-resolved acquisition of three-dimensional (3D) information from the volume of biological material. The concept was initially employed in heterodyne scanning microscopy. OCT has the potential of achieving high-depth resolution, which is determined by the coherence length of the source. This is the length over which a process or a wave maintains strict phase relations; an ideal laser source, for instance, emits light with more than a few kilometers coherence length, while the coherence length of light emitted by a tungsten lamp could be as short as 1 mm. OCT combines interferometry with low-coherence light to produce high-resolution tissue imaging, and it can detect carcinogenesis in epithelial and subepithelial tissues in hamster cheek pouches with an overall sensitivity and specificity of 80%. Newer systems such as Fourier transformed OCT, a complex interferometric optical tomographic system that offers submicrometer resolution, has the potential to give great resolution in a noninvasive way and should yield information about the early changes associated with invasive cancer (Scully et al. 2008).

Ridgway et al. (2006) examined the mucosa of the oral cavity and the oropharynx using OCT in 41 patients during operative endoscopy. OCT imaging was combined with endoscopic photography for gross and histological image correlation. They

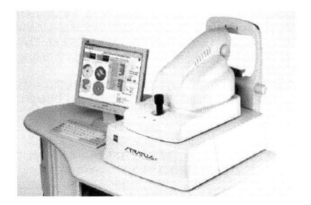

FIGURE 7.8 OCT imaging machine.

found that OCT images of the oral cavity and oropharynx provided microanatomical information on the epithelium, basement membrane, and supporting lamina propria of the mucosa. OCT imaging showed distinct zones of normal, altered, and ablated tissue microstructures for each pathological process studied (Jerjes et al. 2008).

Armstrong et al. (2006) evaluated the ability of OCT to identify the characteristics of laryngeal cancer and measurable changes in the basement membrane, tissue microstructure, and the transition zone at the edge of tumors in 26 OCT examinations. OCT clearly identified basement membrane violation from laryngeal cancer and could identify transition zones at the cancer margin. They suggested that OCT showed potential for assisting in diagnostic assessment (Jerjes et al. 2008).

Wong et al. (2004) performed OCT imaging on 82 patients who underwent surgical endoscopy for various head and neck pathologies. They concluded that OCT has the unique ability to image laryngeal tissue microstructure and can detail microanatomical changes in benign, premalignant, and malignant laryngeal pathologies (Jerjes et al. 2008).

CONCLUSION

Screening and early detection play an important and significant role in treatment as well as the prognosis of cancer. This way of noninvasive diagnosis not only increases the efficiency of early detection but also avoids scalpel injury to the lesion. Hence, promotion of such a noninvasive cancer diagnosis should be encouraged and motivated such that standardized reliability will be achieved.

REFERENCES

Allegra E, Lombardo N, Puzzo L, Garozzo A. The usefulness of toluidine staining as a diagnostic tool for precancerous and cancerous oropharyngeal and oral cavity lesions. *Acta Otorhinolaryngol Ital.* 2009 Aug;29(4):187–190.

Allen JC. Abnormal bleeding response to treatment with toluidine blue and protamine sulphate. *JAMA.* 1949;139:1251–1254.

Armstrong WB, Ridgway JM, Vokes DE, et al. Optical coherence tomography of laryngeal cancer. *Laryngoscope* 2006;116:1107–1113.

Betz CS, Mehlmann M, Rick K, Stepp H, Grevers G, Baumgartner R, Leunig A. Autofluorescence imaging and spectroscopy of normal and malignant mucosa in patients with head and neck cancer. *Lasers Surg. Med.* 1999;25:323–334.

Černý V, Turek Z, Pařízková R. Orthogonal polarization spectral imaging. *Physiol Res.* 2007;56:141–147.

Choo LR, Mansfield JR, Pizzi N, Somorjai L, Jackson M, Halliday WC, Mantsch HH. Infrared spectra of human central nervous system tissue: Diagnosis of Alzheimer's disease by multivariate analyses. *Biospectroscopy* 1995;1:141–148.

Doyle JL, Maiold JH, Weisinger E. Study of glycogen content and "basement membrane" in benign and malignant oral lesions. *Oral Surg.* 1968;26:667–673.

Epstein JB, Oakley C, Millner A, et al. The utility of toluidine blue application as a diagnostic aid in patients previously treated for upper oropharyngeal carcinoma. *OOOOE* 1997;83:537–547.

Epstein JB, Scully C, Spinelli J. Toluidine blue and Lugol's iodine application in the assessment of oral malignant disease and lesions at risk of malignancy. *J Oral Pathol Med.* 1992 Apr;21(4):160–163.

Fagrell B, Intaglieta M. Microcirculation: Its significance in clinical and molecular medicine. *J Intern Med* 1997;241:349–362.

Farah CS, McCullough MJ. A pilot case control study on the efficacy of acetic acid wash and chemiluminescent illumination (ViziLite) in the visualisation of oral mucosal white lesions. *Oral Oncol.* 2007 Sep;43(8):820–824.

Güneri P, Epstein JB, Kaya A, Veral A, Kazandı A, Boyacioglu H. The utility of toluidine blue staining and brush cytology as adjuncts in clinical examination of suspicious oral mucosal lesions. *Int J Oral Maxillofac Surg.* 2011 Feb;40(2):155–161.

Herlin R, Marnay J, Jacob JH, Ollivier JM, et al. A study of the mechanism of staining of the toluidine blue dye test. *Endoscopy.* 1983 Jan;15(1):4–7.

Huang D, Wang J, Lin CP, Puliafito CA, Fujimoto JG. Micron-resolution ranging of cornea anterior chamber by optical reflectometry. *Lasers Surg Med.* 1991;11(5):419–425.

Huber MA. Premalignant lesions. *J Am Dent Assoc.* 2008;139(4):395–396.

Jerjes W, Upile T, Betz C, Abbas S, Sandison A, Hopper C. Detection of oral pathologies using optical coherence tomography. *Eur Oncol.* 2008;4(1):57–59.

Kozlovsky A, Gordon D, Gelernter I, Loesche WJ, Rosenberg M. Correlation between the bana test and oral malodor parameters. *J Dent Res.* 1994 May;73(5):1036–1042.

Lehr HA, Bittinger B, Kirkpatrick CJ. Microcirculatory dysfunction in sepsis: A pathogenetic basis for therapy? *J Pathol* 2000;190:373–386.

Lindeboom JA, Mathura KR, Ince C. Orthogonal polarization spectral (OPS) imaging and topographical characteristics of oral squamous cell carcinoma. *Oral Oncol.* 2006 Jul;42(6):581–585.

Maeda K, Suzuki T, Ooyama Y, Nakakuki K, Yamashiro M, Okada N, Amagasa T. Colorimetric analysis of unstained lesions surrounding oral squamous cell carcinomas and oral potentially malignant disorders using iodine. *Int J Oral Maxillofac Surg.* 2010 May;39(5):486–492.

Maeda K, Yamashiro M, Michi Y, Suzuki T, Ohyama Y, Okada N, Amagasa T. Effective staining method with iodine for leukoplakia and lesions surrounding squamous cell carcinomas of the tongue assessed by colorimetric analysis. *J Med Dent Sci.* 2009 Dec;56(4):123–130.

Manoharan R, Baraga JJ, Feld MS, Rava RP. Quantitative histochemical analysis of human artery using Raman spectroscopy. *J Photochem Photobiol B: Biol.* 1992;16:211–233.

McMahon J, Devine JC, McCaul JA, McLellan DR, Farrow A. Use of Lugol's iodine in the resection of oral and oropharyngeal squamous cell carcinoma. *Br J Oral Maxillofac Surg.* 2010 Mar;48(2):84–87.

Myers EN. The toluidine blue test in lesions of the oral cavity. *CA Cancer J Clin.* 1970 May–Jun;20(3):134–139.

Nagaraju K, Prasad S, Ashok L. Diagnostic efficiency of toluidine blue with Lugol's iodine in oral premalignant and malignant lesions. *Indian J Dent Res.* 2010 Apr–Jun; 21(2):218–223.

Niebel HH, Chomet B. In vivo staining test for the delineation of oral intra-epithelial neoplastic change. *J Am Dent Assoc.* 1964;68:801–806.

Oh ES, Laskin DM. Efficacy of the ViziLite system in the identification of oral lesions. *J Oral Maxillofac Surg.* 2007 Mar;65(3):424–426.

Oliveira AP, Martin AA, Silveira L, Zângaro AR, Zampieri M. Application of principal components analysis to diagnosis hamster oral carcinogenesis: Raman study. *Proceedings of SPIE Vol. 5321, Biomedical Vibrational Spectroscopy and Biohazard Detection Techologies.* Bellingham, WA: SPIE; 2004. doi:10.1117/12.527793.

Olivo M, Bhuvaneswari R, Keogh I. Advances in bio-optical imaging for the diagnosis of early oral cancer. *Pharmaceutics* 2011;3:354–378.

Onizawa K, Saginoya H, Furuya Y, Yoshida H. Fluorescence photography as a diagnostic method for oral cancer. *Cancer Lett.* 1996;108:61–66.

Patton LL, Epstein JB, Kerr AR. Adjunctive techniques for oral cancer examination and lesion diagnosis: A systematic review of the literature. *J Am Dent Assoc.* 2008 Jul;139(7):896–905; quiz 993–994.

Pavlova I, Williams M, El-Naggar A, Richards-Kortum R, Gillenwater A. Understanding the biological basis of autofluorescence imaging for oral cancer detection: High-resolution fluorescence microscopy in viable tissue. *Clin Cancer Res.* 2008 Apr 15;14(8):2396–2404.

Petruzzi M, Lucchese A, Baldoni E, Grassi FR, Serpico R. Use of Lugol's iodine in oral cancer diagnosis: An overview. *Oral Oncol.* 2010 Nov;46(11):811–813.

Poh CF, Zhang L, Anderson DW, Durham JS, Williams M, Priddy RW, et al. Fluorescence visualization detection of field alterations in tumor margins of oral cancer patients. *Clin Cancer Res.* 2006;12:6716–6722.

Rahman F, Tippu SR, Khandelwal S, Girish KL, Manjunath BC, Bhargava A. A study to evaluate the efficacy of toluidine blue and cytology in detecting oral cancer and dysplastic lesions. *Quintessence Int.* 2012 Jan;43(1):51–59.

Ram S, Siar CH. Chemiluminescence as a diagnostic aid in the detection of oral cancer and potentially malignant epithelial lesions. *Int J Oral Maxillofac Surg.* 2005 Jul;34(5):521–527.

Reichart RM. A clinical staining test for the in vivo delineation of dysplasia and carcinoma-in-situ. *Am J Obstetric Gynecology.* 1963;86:703–712.

Richards Kortum R, Svistun E, Alizadeh-Naderi R, El-Naggar A, Jacobs R, Gillenwater A. Vision enhancement system for detection of oral cavity neoplasia based on autofluorescence. *Head Neck.* 2004;26:205–215.

Ridgway JM, Jackson RP, Guo S, et al. In vivo optical coherence tomography of the oral cavity and oropharynx. *Arch Otolaryngol Head Neck Surg.* 2006;132:1074–1108.

Saravanan R, Siar CH. Innovative techniques in the detection of oral cancer and potentially malignant lesions: Chemiluminescence versus tolonium chloride. *Seminar penyelidikan jangka pendek* 2003;34(5):521.

Scully C, Bagan JV, Hopper C, Epstein JB. Oral cancer: Current and future diagnostic techniques. *Am J Dent.* 2008 Aug;21(4):199–209.

Silverman S Jr, Bhargava K, Smith LW, et al. Malignant transformation and natural history of oral leukoplakia in 57,518 industrial workers of Gujarat, India. *Cancer.* 1971;38:1790–1795.

Strong S, Vaughn CW, Incze JS. Toluidine blue in the management of carcinoma of the oral cavity. *Arch Otolaryng.* 1968;87:527–531. [Cited in Mashberg A. Final evaluation of tolonium chloride rinse for screening of high-risk patients with asymptomatic carcinoma. *JADA.* 1983;106:319–323.]

Tooke JE. Microvasculature in diabetes. *Cardiovasc Res.* 1996;32:764–771.

Trullenque-Eriksson A, Muñoz-Corcuera M, Campo-Trapero J, Cano-Sánchez J, Bascones-Martínez A. Analysis of new diagnostic methods in suspicious lesions of the oral mucosa. *Med Oral Patol Oral Cir Bucal.* 2009 May 1;14(5):E210–E216.

Upadhyay J, Rao NN, Upadhyay RB, Agarwal P. Reliability of toluidine blue vital staining in detection of potentially malignant oral lesions—Time to reconsider. *Asian Pac J Cancer Prev.* 2011;12(7):1757–1760.

Vahidy NA, Zaidi SHM, Jafarey NA. Toluidine blue test for detection of carcinoma of the oral cavity: An evaluation. *J Surg Oncol.* 1972;4:434–438. 10.1002/jso.2930040505.

Vercellino V, Gandolfo S, Camoletto D, et al. 1985. Toluidine blue (tolonium chloride) in the early diagnosis of dysplasia's and carcinomas of the oral mucosa. *Minerva Stomatol.* 1985;34:257–261.

Wong BJF, Zhao Y, Yamaguchi M, et al. Imaging the internal structure of the rat cochlea using optical coherence tomography at 0.827 mm and 1.3 mm. *Otolaryngol Head Neck Surg* 2004;130:334–338.

8 Oral Toxicities Due to Cancer Therapy

Balu Karthika

CONTENTS

INTRODUCTION

Oral complications are frequently encountered in patients receiving anticancer therapy, and these complications may result in significant morbidity, treatment delays, dose reductions, and nutritional deficiencies (Ilgenli et al. 2001). Oral cancer and its treatment can cause a variety of problems to patients, including issues with maintaining their daily oral hygiene (Meurman and Gronroos 2010). Smaller and localized tumors have a far lower mortality rate and less morbidity than advanced lesions. Thus, staging of oral cancers is critically important, inasmuch as more advanced tumors require more aggressive therapy. As is to be expected, the more intensive therapeutic approaches used to improve survival also increase the complications. Therefore, preventing or at least minimizing these complications is vital to quality of life and successful rehabilitation (Silverman 1998). Orofacial complications are unfortunately common with all modalities used in the management of patients with malignant disease in the head and neck (Scully and Epstein 1996). Squamous cell carcinoma is the most common cancer in the oral cavity (Surya and Priyanka 2016). Cancer is characterized by increased cell proliferation and diminished apoptosis. The proliferation of atypical cells gives rise to invasive capacity, with the infiltration of body tissues or organs through the bloodstream or lymphatic system—this process is known as *metastasis*. The existing cancer treatments include surgery and radiotherapy, chemotherapy, biological or immune therapy, hormonal therapy, and gene therapy (a form of treatment that is still in the investigational stage), which aim to block cell proliferation (Chaveli-Lopez 2014).

The goal of these regimens is to improve disease control, survival, and quality of life (QoL) through the preservation of function. Recent studies evaluating concomitant chemoradiotherapy (CCRT) regimens suggest small but statistically significant improvements in both disease-free and overall survival. In addition, these regimens offer curative intent to patients with unresectable tumors and thus are encouraging in terms of disease-related outcomes (List et al. 1999). Clinically, fatigue is a symptom commonly defined as a patient's feeling of lack of energy, weariness, or tiredness. About 70% of people with cancer report feelings of fatigue during radiotherapy or chemotherapy, or after surgery (Dios and Leston 2010). This form of fatigue is generally much more disruptive than that associated with other diseases such as depression, multiple sclerosis, or arthritis (Elting et al. 2003). Irrespective of the type of cancer, the related fatigue influences all parts of a patient's QoL and aggravates other distressing symptoms such as pain, nausea, and dyspnea. Fatigue is also a serious problem for people who survive cancer; up to 30% experience the symptom for years after the end of treatment (Lucia et al. 2003). Severe late toxicity after concomitant chemoradiotherapy (CCRT) is common. Older age, advanced T-stage, and larynx/hypopharynx primary site were strong independent risk factors. Neck dissection after CCRT was associated with an increased risk of complications (Machtay et al. 2008).

CANCER THERAPY ORAL TOXICITIES

Contemporary cancer treatment modalities commonly include surgical resection, chemotherapy, radiotherapy, and hematopoietic stem cell transplantation (HSCT), a form of immunotherapy, either administered alone or used in combination. Although the effectiveness of cancer treatment has continued to improve over the past decades, collateral damage to the head and neck structures is frequently encountered as an unwanted consequence. Radio- and chemotherapy can cause direct harm to the soft and hard tissues of the oral structures, whereas their systemic toxicity can give rise to indirect damages. These oral complications, be they acute or chronic, may arise throughout and after cancer treatment and often encompass mucositis, dysgeusia, and infectious diseases (Wong 2014). Cancers that invade the head and neck area can cause physical barriers and disruption to the process of eating and swallowing. Neurological complications, from cancers that affect the nervous system pathways (often caused by compression of critical nerve pathways by tumor masses) may result in reduced or absent ability to swallow (dysphagia). Dysphagia may be complicated by ulcerous sores in the oral cavity (stomatitis) or through the gastrointestinal tract (mucositis) (Broadfield and Hamilton 2006). In head and neck cancer, quality of survival is critically influenced by performance, or functional ability, in areas of eating and speaking (List et al. 1990). Curative treatment with CCRT for patients with advanced cancer of the head and neck has adverse effects on many functions of the upper respiratory and digestive systems. Sequelae such as pain, edema, xerostomia, and fibrosis negatively affect mouth opening (trismus), chewing, swallowing, and speech (van der Molen et al. 2011). Oral cancer and particularly its treatment can cause problems for the daily maintenance of oral health.

Surgical treatment mutilates tissue anatomy, and radiotherapy may cause mucositis, tissue constriction, and irreversible damage to salivary glands with subsequent dry mouth. Cytostatic drugs affect both the local and systemic defensive factors leading easily to persistent or masked infections. Hyposalivation and xerostomia not only affect dental health but also burden the patient with oral dryness or mucosal pain, reduce taste and smell, increase the risk for dental and mucosal infections, and cause problems of speaking and mastication, thus decreasing the QoL (Dios and Leston 2010). Oral mucositis is a common toxicity associated with both antineoplastic head and neck radiation and chemotherapy (Sonis 2004). It has been hypothesized that mucositis leads to infection and bleeding through interruption of the mucosa (Elting et al. 2003). Reconstructive surgery together with prosthetic and dental rehabilitation causes further problems to the patient and a frequent need to modify treatment plans by the professional oral health team. Consequently, guidelines and recommendations for treating oral health problems of oral cancer patients have been presented, especially for preventing infections and other dental diseases (Meurman and Gronroos 2010). Pain may be the initial symptom in oral cancer and is a common complaint both in patients awaiting treatment and in those already in treatment. Cancer pain may be due to tumor progression, invasive diagnostic or therapeutic procedures, toxicity of chemotherapy and radiation therapy, infection, or muscle aches when patients limit physical activity. Several factors make head and neck cancer pain management particularly difficult: the erosive nature of tumors located in this region; the rich innervation of the head and neck; the "dynamic pain" provoked by functional movements such as swallowing, talking, or chewing; and the neuropathic pain induced by chemotherapy and radiation therapy (Dios and Leston 2010).

Frequencies of oral complications from treatment may vary, depending on the type of therapy given; some frequency estimates include the following:

- 10% adjunctive chemotherapy for solid tumors (low risk)
- 40% primary chemotherapy (e.g., for hematologic malignancies) (intermediate risk)
- 80% hematopoietic stem cell transplantation (HSCT) (high risk)
- 100% head and neck radiation therapy to fields involving the oral cavity

The most common oral complications of cancer therapy include mucositis, infection (local or systemic), salivary gland dysfunction, taste alteration, and pain, which result directly or indirectly from the side effects of therapy. These complications can lead to secondary complications such as nutritional disorder, xerostomia, or hemorrhage that may not be resolved with aggressive medical, nursing, and dental interventions. As much as possible, oral complications should be prevented using good oral hygiene techniques (Broadfield and Hamilton 2006).

Although more patients are being cured of their disease through therapies, a substantial percentage of survivors suffer from significant treatment-related adverse effects. Head and neck cancer and its treatment can affect both disease-specific health-related quality of life (HRQoL; e.g., salivary and swallowing functions) and the more general domains of HRQoL, such as physical, mental, and social health (Langendijk et al. 2008). A frequent complication of anticancer treatment, oral and gastrointestinal (GI) mucositis, threatens the effectiveness of therapy because it leads to dose reductions, increases health-care costs, and impairs patients' QoL (Sonis et al. 2004). Malnutrition is a frequent complication in patients with cancer and can negatively affect the outcome of treatments. Side effects of anticancer therapies can also lead to inadequate nutrient intake and subsequent malnutrition (Santarpia et al. 2011).

Cancer patients treated with traditional surgery with or without radiation therapy have documented significant and extensive impairment, including disfigurement, speech disorder, dry mouth, stiffening/constriction of local tissues, chewing and swallowing dysfunction, mood disturbance, anxiety, and depression (List et al. 1999).

ORAL COMPLICATIONS DUE TO NONSURGICAL CANCER THERAPY

CCRT is a standard treatment for patients with locally advanced head and neck squamous cell carcinoma (HNSCC) treated nonsurgically. Meta-analyses show an improved 5-year survival by approximately 8% when CCRT is compared with radiation therapy alone. The advantage of this approach with respect to disease-free survival and locoregional control is greater than 8% (Machtay et al. 2008). Aggressive treatments of malignant diseases may produce unavoidable toxicities to normal cells. The oral cavity is susceptible to direct and indirect toxic effects of cancer chemotherapy, ionizing radiation, and HSCT (Deshpande and Dhokar 2015). While there are undisputed advantages to CCRT for locoregional control, it increases toxicity when compared with radiation therapy alone (Machtay et al. 2008). These complications may include mucositis, xerostomia, dental caries, loss of taste, trismus, infection, osteoradionecrosis, and abnormalities of growth and development. Preventing and treating oral complications of cancer therapy involve identifying the patient at risk, starting preventive measures before cancer therapy begins, and treating complications as soon as they appear. These patients can visit a dentist for management of their complications (Deshpande and Dhokar 2015).

In both chemotherapy- and radiation-associated mucositis, pain intensity is related to the extent of tissue damage and the degree of local inflammation. Management involves the aggressive use of analgesics, often opioids. There is no evidence that patient-controlled analgesia is better than a continuous infusion method for controlling pain, but less opiates are used per hour, and pain duration is shorter. Polyvinylpyrrolidone–sodium hyaluronate shields the exposed nerve endings that give rise to overstimulation, reducing the pain and disruption of the oral mucosa. Laser therapy may also be effective in preventing and treating oral pain related to radiation therapy- and chemotherapy-induced mucositis (Dios and Leston 2010). Organ preservation with radiotherapy and concomitant chemotherapy has become an accepted treatment modality in advanced head and neck cancer. Unfortunately, organ preservation is not synonymous with function preservation (van der Molen et al. 2009).

ORAL COMPLICATIONS DUE TO RADIATION THERAPY

Late radiation-induced toxicity, particularly swallowing and xerostomia, has a significant impact on the more general dimensions of health-related QoL. These findings suggest that the development of new radiation-induced delivery techniques should not only focus on reduction of the dose to the salivary glands, but also on anatomic structures that are involved in swallowing (Montazeri 2008).

Radiation therapy injures the parenchyma of the salivary gland, and eventually this leads to fibrosis and secretary hypofunction. The effects are dose related and permanent, resulting in a condition known as postirradiation xerostomia (Johnson et al. 1993). Intensity-modulated radiation therapy (IMRT) is a treatment planning system designed to decrease toxicity by limiting the radiation dose to critical nerve and salivary structures. However, although designed to reduce damage to nerve and salivary structures, this planning modality may actually increase the total radiation dose given to other structures not shielded in the treatment planning algorithm, resulting in increased skin and mucosal toxicity (Givens et al. 2009).

ORAL COMPLICATIONS DUE TO CHEMOTHERAPY

Oral lesions are a common complication in patients with cancer who are receiving chemotherapy. These events are probably the result of the direct effect of cytotoxic drugs in the rapidly dividing oral epithelium and are manifested by thinning and ulceration of the mucosa. They may be indirectly due to chemotherapy-induced myelosuppression seen as oral bleeding and local infections. These treatment-associated oral complications, particularly mucositis, may produce severe discomfort and pain, which interfere with oral feeding, delays or dosage limitations of antineoplastic treatment, and in some patients life-threatening septicemia (Ramirez-Amador et al. 1996).

Mucositis induced by antineoplastic drugs is an important, dose-limiting and costly side effect of cancer therapy. The ulcerative lesions produced by stomato-toxic chemotherapy are painful, restrict oral intake, and importantly, act as sites of secondary infection and portals of entry for the endogenous oral flora (Sonis 1998). Mucositis is a common cause of morbidity during chemotherapy. The incidence of National Cancer Institute (NCI) grade 3–4 oral and GI mucositis derived from clinical trials of standard-dose chemotherapy is estimated between 5% and 15%. Chemotherapy with 5-fluorouracil (5-FU) or irinotecan is associated with rates of oral or GI mucositis exceeding 15% (Dios and Leston 2010). Among patients receiving high-dose regimens or concomitant radiotherapy to the head, neck, thorax, chest wall, abdomen, or pelvis, rates may exceed 40% and significantly affect QoL. However, rates of mucositis after most chemotherapy regimens administered to patients with solid tumors probably are far lower (Elting et al. 2003).

Oral mucositis can be very painful and can significantly affect nutritional intake, mouth care, and QoL. For patients receiving high-dose chemotherapy prior to hematopoietic cell transplantation, oral mucositis has been reported to be the most debilitating complication of transplantation. Infections associated with the oral mucositis lesions can cause life-threatening systemic sepsis during periods of profound immunosuppression. Moderate to severe oral mucositis has been correlated with systemic infection and transplant-related mortality. In patients with hematologic malignancies receiving allogeneic hematopoietic cell transplantation, increased severity of oral mucositis was found to be significantly associated with an increased number of days requiring total parenteral nutrition and parenteral narcotic therapy, increased number of days with fever, increased incidence of significant infection, increased time in hospital, and increased total inpatient charges. In patients receiving chemotherapy for solid tumors or lymphoma, the rate of infection during cycles with mucositis was more than twice that during cycles without mucositis and was directly proportional to the severity of mucositis. Infection-related deaths were also more common during cycles with both oral and GI mucositis. In addition, the average duration of hospitalization was significantly longer during chemotherapy cycles with mucositis. Importantly, a reduction in the next dose of chemotherapy was twice as common after cycles with mucositis than after cycles without mucositis. Thus, mucositis can be a dose-limiting toxicity of cancer chemotherapy with direct effects on patient survival (Lalla et al. 2008).

According to the World Health Organization guidelines for patients with pain of moderate severity or greater, opioid analgesics are the mainstay of cancer pain management. The management of excessive adverse effects remains a major clinical challenge (Chemy et al. 2001).

ORAL COMPLICATIONS DUE TO SURGERY

Functional impairment is common in advanced cancer patients and has the capacity to engender significant psychosocial morbidity (Cheville 2001). Patients with head and neck cancer suffer from functional impairments due to intense treatment (Ahlberg et al. 2011). Surgical procedures are useful in selected patients to debulk tumors and thus reduce the symptoms of obstruction or compression. Pain control is usually a secondary goal when curative tumor resection is performed, whereas it is usually the operative indication in palliative surgery for unresectable tumors. Surgery may leave patients with major changes in anatomy and functionality that require further rehabilitation and continued pain management. Careful surgical technique can reduce the severity of postoperative pain. Gentle tissue handling, use of nerve- and vessel-sparing procedures, avoidance of tissue ischemia, careful neurolysis, and selection of non-muscle-splitting incisions can contribute to less painful surgery and recovery. In the immediate postoperative period, the whole pain-control armamentarium may be used. In patients with oral cancer, mandibular resection is frequently unavoidable in order to obtain disease-free margins. Patients undergoing segmental resection usually report more pain than those in whom rim resection is performed, probably because the choice of surgical technique is determined by the size of the initial lesion. Postoperative pain scores after segmental mandibulectomy and reconstruction using composite free tissue transfer were relatively good, with little difference compared with rim resections. In patients in whom prosthetic rehabilitation was performed following partial jaw resection, it was demonstrated that patients with maxillary resection showed higher pain threshold values than patients with segmental mandibulectomy. In patients undergoing major surgical procedures including myocutaneous flap reconstruction, both epidural and intravenous morphine appear to provide good pain relief. Postoperative pain usually decreases progressively over the subsequent year; the poorest recovery values are found in patients diagnosed with advanced stages of the disease. Despite the fact that many adverse effects of cancer treatment are now well controlled, there are some, such as mucositis and salivary gland hypofunction, that continue to be almost inevitable outcomes of oral cancer treatment. Compared to other toxicities that arise during treatment for head and neck cancer, mucositis is conceivably the most debilitating and painful. Mucositis-induced treatment interruptions and dose reductions have negative consequences on the outcome of treatment of head and neck cancer (Dios and Leston 2010). Swallowing problems are considered to be the most prominent symptom after treatment, and lead to weight loss and probably also reduced health-related QoL. Another important loss of function is stiffness and pain in the neck and shoulders (Ahlberg et al. 2011).

ASSESSMENT OF ORAL TOXICITIES

The Performance Status Scale for Head and Neck Cancer Patients was developed in consultation with experts in the fields of otolaryngology, surgical oncology, and speech and swallowing rehabilitation science. The scale was designed to measure the unique disabilities of head and neck cancer patients in areas of eating and speaking. Surgery and/or adjuvant therapies often result in cosmetic and functional deficiencies. Facial disfigurement and disturbances of speech and eating are the most obvious handicaps. Voice loss and facial distortion are traumatic. These disabilities are disfiguring and may affect both self-esteem and the reactions of others. Personal relationships and employment opportunities may be altered. Similarly, the eating patterns of head and neck patients become less socially acceptable. Increased time for feeding is often required, considerable messiness is involved, and special food preparation is needed. Related problems associated with eating include nutrition management, oral inconsistency (inability to control saliva), and prosthodontics and dental care. In addition to speaking and eating problems, the social involvements of head and neck cancer patients are often curtailed. Communication difficulties may result in frustration, anxiety, and isolation from others. Similarly, problems with eating and the ensuing embarrassment may interfere with the patient's socializing with others in situations where food or drinks are a significant aspect of the activity (List et al. 1990).

Table 8.1 illustrates rehabilitation-related issues for cancer treatment in management phases of cancer (Gerber 2001).

TABLE 8.1
Rehabilitation-Related Issues for Cancer Patients

Phase of Cancer	Patient Needs	Symptoms	Impact of Symptoms on Function
I. Pretreatment and evaluation	Information about treatment options and impact of illness	Pain Anxiety Depression	Daily routines Sleep/fatigue
II. Treatment	Information Support Rehabilitation interventions Help with daily routines Vocational home, etc.	Pain Anxiety Loss of mobility Wound/skin care Speech/swallowing	Daily routines Sleep/stamina Self-care Cosmesis Communication
III. Posttreatment	Support Rehabilitation intervention	Plain/weakness Anxiety/depression Loss of mobility Edema Fatigue/stamina	Sleep/fatigue Activities of daily living Vocational/avocational Cosmesis
IV. Recurrence	Education Support Rehabilitation intervention	Pain/weakness Anxiety/depression Fatigue/stamina Edema Bony instability Anorexia	Sleep/fatigue Disability Disruption of routines Cosmesis Vocational/avocational
V. End of life	Education Support Palliative rehabilitation	Pain Fatigue Anorexia	Dependence Immobility

CONCLUSION

The performance status of head and neck cancer patients is an important consideration in the selection and evaluation of treatment options and outcomes, and rehabilitation strategies. Traditionally, disease control and length of survival have been the measures of treatment efficacy. Today, however, with increasing survival rates, the psychosocial dimensions of cancer care and the quality of survival have become areas of concern (List et al. 1990). Prognosis, however, is still ultimately dependent on the clinical stage of the tumor at the initial diagnosis with respect to size, depth, extent, and metastasis as recurrence rates and survival rates plummet with disease progression (Deng et al. 2011). It has been shown that assessing QoL in cancer patients could contribute to improved treatment and could even be as prognostic as medical factors could be prognostic (Johnson et al. 1993).

REFERENCES

Ahlberg A, Engstrom T, Nikolaidis P, Gunnarsson K, Johansson H, Sharp L, Laurell G. Early self-care rehabilitation of head and neck cancer patients. *Acta Oto-Laryngologica*. 2011;131:5;552–561.

Broadfield L, Hamilton J. Best practice guidelines for the management of oral complications from cancer therapy. In: *Best Practise Guidelines for the Management of Complications from Cancer Therapy.* Halifax, Nova Scotia, Canada: Supportive Care Cancer Site Team, Cancer Care Nova Scotia; 2006;1–104.

Chaveli-Lopez B. Oral toxicity produced by chemotherapy: A systemic view. *J Clin Exp Dent.* 2014;6(1):e81–e90.

Cherny N, Ripamonti C, Pereira J, Davis C, Fallon M, McQuay H, et al.; for the expert working group of the European Association of Palliative Care Network. Strategies to manage the adverse effects of oral morphine: An evidence based report. *J Clin Oncol.* 2001;19:2542–2554.

Cheville A. Rehabilitation of patients with advanced cancer. *Cancer.* 2001;92(4 Suppl):1039–1048.

Deng H, Sambrook PJ, Logan RM. The treatment of oral cancer: An overview for dental professionals. *Aust Dent J.* 2011;56:244–252.

Deshpande TS, Dhokar A. Oral complications of non-surgical cancer therapies. *Int J Prev Clin Dent Res.* 2015;2(5):52–57.

Dios PD, Leston JS. Oral cancer pain. *Oral Oncol.* 2010;46:448–451.

Elting LS, Cooksley C, Chambers M, Cantor SB, Manzullo E, Rubenstein EB. The burdens of cancer therapy. Clinical and economic outcomes of chemotherapy-induced mucositis. *Cancer.* 2003;98:1531–1539.

Gerber LH. Cancer rehabilitation into the future. *Cancer.* 2001;92(4 Suppl):975–979.

Givens DJ, Karnell LH, Gupta AK, Clamon GH, Pagedar NA, Chang KE, et al. Adverse events associated with concurrent chemoradiation therapy in patients with head and neck cancer. *Arch Otolaryngol Head Neck Surg.* 2009;135(12):1209–1217.

Ilgenli T, Oren H, Uysal K. The acute effects of chemotherapy upon the oral cavity: Prevention and management. *Turkish J Cancer.* 2001;31(3):93–105.

Johnson JT, Ferreti GA, Nethery WJ, Vladez IH, Fox PC, Ng D, et al. Oral philocarpine for post-irritation xerostomia in patients with head and neck cancer. *N Engl J Med.* 1993;329:390–395.

Lalla RV, Sonis ST, Peterson DE. Management of oral mucositis in patients with cancer. *Dent Clin North Am.* 2008;52(1):61–viii:1–17.

Langendijk JA, Doornaert P, Verdonck de Leeuw IM, Leemans CR, Aaronson NK, Slotman BJ. Impact of late treatment related toxicity on quality of life among patients with head and neck cancer treated with radiotherapy. *J Clin Oncol.* 2008;26:3770–3776.

List MA, Ritter-Sterr C, Lansky SB. A performance status scale for head and neck cancer patient. *Cancer.* 1990;66:564–569.

List MA, Siston A, Haraf D, Schumm P, Kies M, Stenson K, Vokes EE. A prospective examination. *J Clin Oncol.* 1999;17(3):1020–1028.

Lucia A, Earnest C, Perez M. Cancer-related fatigue: Can experience physiology assist oncologist? *The Lancet Oncol.* 2003;4:616–625.

Machtay M, Moughan J, Trotti A, Garden AS, Weber RS, Cooper JS, Forastiere A, Ang KK. Factor associated with severe late toxicity after concurrent chemoradiation for locally advanced head and neck cancer: An RTOG analysis. *J Clin Oncol.* 2008;26:3582–3589.

Meurman JH, Gronroos L. Oral and dental health care of oral cancer patients: Hyposalivation, caries and infections. *Oral Oncol.* 2010;46:464–467.

Montazeri A. Health related quality of life in breast cancer patients: A bibliographic review of the literature from 1974 to 2007. *J Exp Clin Cancer Res.* 2008;27:1–31.

Ramirez-Amador V, Esquivel-Pedraza L, Mohar A, Reynoso-Gomez E, Volkow-Fernandez P, Guarner J, Sanchez-Mejorda G. Chemotherapy associated oral mucosal lesion in patients with leukemia or lymphoma. *Oral Oncol Eur F Cancer.* 1996;32B(5):322–327.

Santarpia L, Contaldo F, Pasanisi F. Nutritional screening and early treatment of malnutrition in cancer patients. *J Cachexia Sarcopenia Muscle.* 2011;2:27–35.

Scully C, Epstein JB. Oral health care for the cancer patient. *Oral Oncol Eur F Cancer.* 1996;32B(5):281–292.

Silverman S, Jr. Oral cancer. *Oral Surg Oral Med Oral Pathol Oral Radiol Endodontol.* 1998;88:122–126.

Sonis ST. Mucositis as a biological process: A new hypothesis for the development of chemotherapy induced stomatotoxicity. *Oral Oncol.* 1998;34:39–43.

Sonis ST. A biological approach to mucositis. *J Support Oncol.* 2004;2:21–36.

Sonis ST, Elting LS, Keefe D, Peterson DE, Schubert M, Hauer-Jensen M, et al. Prospectives on cancer therapy-induced mucosal injury. *Cancer.* 2004;100(9 Suppl):1995–2025.

Surya V, Priyanka V. Serum biomarkers: Evaluation of serum albumin and serum albumin:globulin ratio in oral leukoplakia and oral squamous cell carcinoma. *J Dental Oral Health.* 2016;2:2–6.

van der Molen L, van Rossum MA, Burkhead LM, Smeele LE, Rasch CRN, Hilgers FJM. A randomized preventive rehabilitation trial in advanced head and neck cancer patients treated with chemoradiotherapy: Feasibility, compliance and short term effects. *Dysphagia.* 2011;26:155–170.

van der Molen L, van Rossum MA, Burkhead LM, Smeele LE, Hilgers FJM. Functional outcomes and rehabilitation strategies in patient treated with chemoradiotherapy for advanced head and neck cancer: A systematic review. *Eur Arch Oto-Rhino-Laryngol.* 2009;266:889–900.

Wong HM. Oral complication and management strategies for patient undergoing cancer therapy. *Sci World J.* 2014;1–14.

9 Herbal Therapy for Cancer
Clinical and Experimental Perspectives

Devaraj Ezhilarasan

CONTENTS

INTRODUCTION

Cancer

Cancer is the uncontrolled growth of cells, which can invade and spread to distant sites of the body. It is one of the major and significant public health problems worldwide and is the second leading cause of death in the United States (Siegel et al. 2016). The International Agency for Research on Cancer (IARC) estimated the incidence of mortality and prevalence from major types of cancer, at a national level, for 184 countries of the world and revealed that there were 14.1 million new cancer cases, 8.2 million cancer deaths, and 32.6 million people living with cancer (within 5 years of diagnosis) in 2012 worldwide (GLOBOCAN 2012). The statistics, based on GLOBOCAN worldwide estimates, show the most commonly diagnosed cancers were lung (1.82 million), breast (1.67 million), and colorectal (1.36 million); the most common causes of cancer death were lung cancer (1.6 million deaths), liver cancer

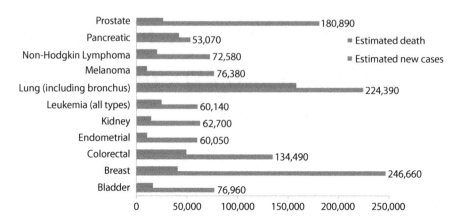

FIGURE 9.1 Estimated numbers of new cases and deaths for each common cancer type in the United States. (From National Cancer Institute. Available from: https://www.cancer.gov/types/common-cancers. Accessed November 28, 2016.)

(745,000 deaths), and stomach cancer (723,000 deaths) (Ferlay et al. 2015). It was projected that by 2030 there will be 26 million new cancer cases and 17 million cancer deaths per year (Thun et al. 2010). Among the different cancer types, lung cancer is reported as one of the deadly malignancies, and also the most common cause of death from cancer. The American Cancer Society estimated cancer-associated deaths and new cases in the United States alone and suggested that more individuals are likely to die of lung, breast, colon, and prostate cancers combined. The most common type of cancer on the list is lung cancer, with more than 224,390 new cases expected in the United States in 2016 (Figure 9.1).

Different Treatment Approach

It is estimated that over 25% of the population of the United States will face a diagnosis of cancer during their lifetime, with more than 1.6 million new cancer patients diagnosed each year. Less than a quarter of these patients will be cured solely by surgery and/or local radiation. Most of the remainder will receive systemic chemotherapy at some time during their illness. Despite considerable efforts and development of various treatment approaches, cancer remains an aggressive disease worldwide. However, early detection, accurate diagnosis, and effective treatment, including pain relief and palliative care, help to increase cancer survival rates and reduce suffering. Treatment options include surgery, chemotherapy, and radiotherapy, tailored to tumor stage, type, and available resources (World Health Organization [WHO] 2016a).

More than 30% of cancer deaths could be prevented by modifying or avoiding key risk factors, for instance, avoiding consumption of tobacco in case of lung and oral cancers. Cancer patients are currently being treated with chemotherapy, palliative therapy, radiotherapy, and surgical removal of tumor based on their cancer types and stage. Cancer chemotherapy is given for cancer patients to prevent the progression of cancer cell growth. However, the synthetic chemotherapeutic drugs given to cancer patients have several untoward effects, for instance, nausea, vomiting, and dizziness (Carelle et al. 2002). Radiotherapy, a frequent mode of cancer treatment, is often restricted by dose-related toxicity and development of therapeutic resistance and significant adverse effects. Moreover, novel synthetic chemotherapeutic agents currently in use clinically have not succeeded in fulfilling expectations despite the considerable cost of their development. Therefore, comprehensive cancer control plans are needed to improve cancer prevention and care, especially in low- and middle-income countries. Despite developments in diagnosis and advancement of chemotherapy, surgery, and radiation, along with various palliative treatments, cancer remains a great challenge for clinical therapy. Currently, complementary and alternative medicines provide treatments for various cancers, especially using herbal medicine derived from medicinal plants.

COMPLEMENTARY ALTERNATIVE MEDICINE

According to the WHO, the terms *complementary medicine* or *alternative medicine* are used interchangeably with traditional medicine in some countries. They refer to a broad set of health-care practices that are not part of that country's own tradition and are not integrated into the dominant health-care system (WHO 2016a). The role of complementary and alternative medicine (CAM) treatments in oncology have always been subjected to matter of discussion. In spite of that, it is estimated that about half of cancer patients experience at least one form of CAM in their life and because of the growing spread of these on the Internet, the proportion is destined to grow. Over the last three decades, these alternative medicines have come to clinical approaches such as massage, herbal remedies, and chiropractic treatments that had not previously been considered components of alternative medicine. These practices have entered mainstream society and culture; surveys confirm that 30%–40% of the U.S. public uses alternative medicine so defined (Burstein et al. 1999).

It was reported in a 2007 National Health Interview Survey, that 83 million U.S. adults spent $33.9 billion out of pocket on visits to CAM practitioners and on purchases of CAM products, classes, and materials. In total, there were approximately 354 million visits to CAM practitioners and approximately 835 million purchases. These data show the increasing popularity and advocacy for the use of CAM among patients. It has also been said that about one in four patients with prostate cancer using at least one CAM method (Wilkinson et al. 2008). The mechanisms of beneficial preventive and therapeutic effects achieved by traditional and complementary medicine are currently being deciphered in molecular medicine, and the role of CAM treatments in oncology has always been heavily debated. However, recent studies have been focused on herbal medicines as potent anticancer drug candidates (Bozza et al. 2015). Subsequently, new strategies are evolving to control and treat cancer. One such strategy could be the use of herbal plant-derived compounds alone or along with standard chemotherapeutic drugs as adjuvants.

HERBAL MEDICINE

Plant substances are a large and diverse group of compounds that are found naturally in fruits, vegetables, spices, and medicinal plants. These medicinal plants have been used for medical purposes since the beginning of human history and are the basis of modern medicine. Herbal medicines include herbs, herbal materials, herbal preparations, and finished herbal products that contain as active ingredients parts of plants, or other plant parts, or combinations. The traditional use of herbal medicines refers to the long historical use of these medicines. Their use is well established and widely acknowledged to be safe and effective and may be accepted by national authorities. Active ingredients refer to phytochemical of herbal medicines with therapeutic activity. In herbal medicines where these phytochemicals have been identified, the preparation of herbal medicines should be standardized to contain a defined amount of the active ingredients, if adequate analytical methods are available. In cases where it is not possible to identify the active ingredients, the whole herbal medicine may be considered as one active ingredient (WHO 2016b).

It has been said that more than 3,000 plant species have been documented to treat cancer and about 30 plant-derived compounds have been isolated so far and have been tested in cancer clinical trials. Most chemotherapeutic drugs for cancer treatment are molecules identified and isolated from plants or their synthetic derivatives. For instance, vinblastine/vincristine (Oncovin) is derived from a medicinal plant (*Vinca rosea*) and are common chemotherapy drugs used to treat many cancers, including leukemia, lymphoma, myeloma, breast cancer, and head and neck cancer. Our hypothesis was that whole plant extracts selected according to ethnobotanical sources of historical use might contain multiple molecules with antitumor activities that could be very effective in killing human cancer cells. The aim of this chapter is to present the current state of knowledge concerning the molecular targets of tumor angiogenesis and the plant-derived active substances such as polyphenols (resveratrol, curcumin, and genistein), alkaloids (berberine), phytohormones, carbohydrates, and terpenes, and their activity against cancer.

PLANT-DERIVED ACTIVE COMPOUNDS

CURCUMIN

Curcumin (diferuloylmethane) (Figure 9.2) is a substance obtained from the root of the turmeric plant *Curcuma longa*, which has the feature of being a yellow or orange pigment. It is also the main component of curry powder commonly used throughout South Asia for its flavor and medicinal properties. Curcumin's (CUR) pharmacological properties are that it slows or reverses cellular proliferation and enhances apoptosis and differentiation associated with a diverse array of molecular effects.

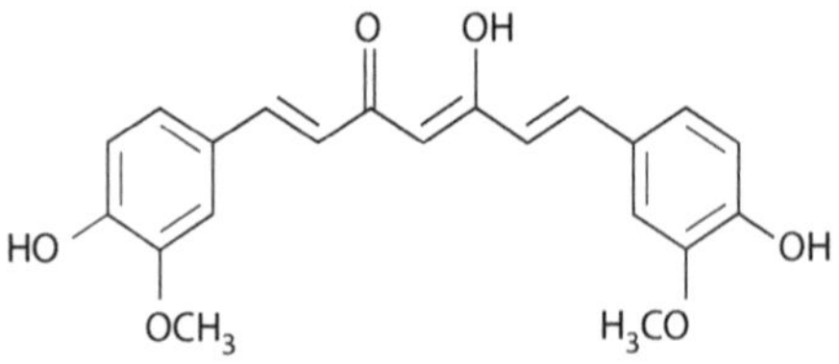

FIGURE 9.2   Chemical structure of curcumin ($C_{21}H_{20}O_6$; CAS No. 458-37-7).

These properties make CUR a leading chemopreventive agent (Unlu et al. 2016). This compound has been studied *in vivo* and *in vitro* for its myriad biological effects against oxidative stress, inflammation, microbial infections, and cancer (Pavan et al. 2016). Here we review the anticancer efficacy of CUR studied in experimental and human subjects.

Breast cancer is among the most common malignant tumors. It is the second leading cause of cancer mortality among women in the United States. Curcumin has been studied against different human breast cancer cell lines as the luminal MCF-7 and the basal-like MDA-MB-231. These studies compare the anticancer efficacy of CUR with the standard anticancer drug—paclitaxel. When these two compounds were used alone or in combination, they significantly inhibited B-cell lymphoma-2 (Bcl-2, anti-apoptotic protein) expression and enhanced Bax protein (pro-apoptotic protein) expressions. These activities of CUR corroborated with the apoptotic effect. This study sheds light that CUR can be considered in the synergistic therapy of breast cancer reducing the associated toxicity with the use of drugs (Muthoosamy et al. 2016; Quispe-Soto and Calaf 2016). Interestingly, CUR inhibits migration and the invasive properties (metastasis) through epithelial-mesenchymal transition (EMT) in breast cancer cell lines (Gallardo and Calaf 2016).

Non-small cell lung cancer (NSCLC) therapy is a challenge owing to poor prognosis and a low survival rate. In a study, Bax expression was increased while the expression of Bcl-2 and B-cell lymphoma-xL (Bcl-xL) was decreased by CUR in small-cell lung cancer and thus, induced apoptosis accompanied by increasing intracellular reactive oxygen species (ROS) levels. Mitochondrial membrane potential (MMP) was decreased, and in turn, the release of cytochrome c into the cytosol was induced, and then caspase-9 and caspase-3 were activated (Yang et al. 2012). The pro-apoptotic potential of CUR is shown in a dose- and time-dependent manner in human lung cancer cells. In addition, CUR treatment caused an upregulation of Bax and Bad, an increased level of ROS accompanied endoplasmic reticulum (ER) stress in these cells. These alterations reduce the MMP consequent to caspase-3 activation. The activation of the extrinsic pathway through increased FAS/CD95 expression promotes caspase-8 activation, and cells were found to be arrested at G2/M cell cycle phase after CUR treatment. These data are confirmed by using a caspase-8 inhibitor, which decreased the apoptosis in these cells (Wu et al. 2010). A recent study evaluated the therapeutic efficacy of CUR (as a nanoformulation) on a NSCLC xenograft model. Xenograft tumors in nude mice were treated with 20 mg/kg subcutaneous injection of CUR. This study highlights the decreased expression of Ki-67, one of the cell proliferation markers (Ranjan et al. 2016). In addition, CUR could inhibit Janus kinase2 (JAK2) activity and reduce tumor spheres via inhibiting the JAK2/signal transducer and activator of transcription 3 (STAT3) signaling pathway, playing a key role in many cellular processes such as cell growth. Thus, CUR strongly repressed tumor growth in the lung cancer xenograft nude mouse model (Wu et al. 2015).

Hepatocellular carcinoma, one of the most common cancers worldwide, is reported to feature relatively high morbidity and mortality. Curcumin has been studied against diethylnitrosamine (DEN)-induced hepatocarcinogenesis in rats. The DEN-induced liver distortions were significantly mitigated by CUR. Curcumin remarkably suppressed the serum levels of α-feto-protein, interleukin-2 (IL-2), interleukin-6 (IL-6), alanine transaminases (ALT), and malondialdehyde, as well as gene expression of IL-2 and IL-6, and increased the gene expression and enzymatic activities of glutathione peroxidase, glutathione reductase, catalase, and super oxide dismutase (Kadasa et al. 2015). In another study, using the same DEN model of hepatocarcinogenesis, the overexpression of the angiogenic and anti-apoptotic factors, transforming growth factor-β (TGF-β), and protein kinase B (PKB) were reduced, while caspase-3 expression was improved by CUR treatment. Liver marker enzymes such as aspartate transaminases (AST) and ALT and oxidative stress were normalized after CUR treatment (Abouzied et al. 2015). Degenerative and apoptotic changes were reported in CUR-treated hepatoma cells (Abdel-Lateef et al. 2016). In addition, CUR treatment was reported to inhibit the growth of liver cancer in a dose-dependent manner in nude mice (Dai et al. 2013).

When colon cancer cells were treated with CUR and silymarin (plant-derived flavonolignan) together, a synergistic anticancer effect was observed (Montgomery et al. 2016). β-Catenin is responsible for the expression of cell proliferation/or cell cycle-related target genes, such as c-myc and cyclin D1, was regulated by the Wnt signaling pathways, and plays a pivotal role in the regulation of cell proliferation in colorectal cancer (CRC) (Chung et al. 2013). For these reasons, studies targeted CUR on β-catenin. Curcumin and its derivatives suppressed β-catenin response transcription without altering the intracellular level of β-catenin, thereby reducing the expression of Wnt/β-catenin pathway transcriptional coactivator p300 and inhibiting β-catenin-related transcription, thus playing a inhibitory role in CRC cell growth and their proliferation (Ryu et al. 2008).

Cervical cancer is one of the most common cancers in women worldwide, and it is a prominent cause of cancer mortality. Curcumin has been shown to induce cytotoxic cell death via induction of DNA damage, and chromatin condensation *in vitro* in HeLa human cervical cancer cells (Shang et al. 2016). A combination of CUR and ellagic acid (a natural phenol antioxidant found in numerous fruits and vegetables) reported to synergistically induce ROS generation, DNA damage, p53 accumulation, and apoptosis in HeLa cervical carcinoma cells (Kumar et al. 2016). This compound has also been tested against several other cervical cancer cells (C33A, CaSki, HeLa, and ME180) concurrently with normal epithelial cells. In this study, CUR elevated ROS levels in different cervical cancer cells, but not in normal epithelial cells. Further, CUR promotes endoplasmic reticulum (ER) stress-mediated apoptosis only in cervical cancer cells through ROS generation. Curcumin was found to inhibit proliferation and migration in SiHa and HeLa cells. TGF-β activates the Wnt/β-catenin signaling pathway in cancerous HeLa cells, and CUR has been found to downregulate this pathway by inhibiting β-catenin (Thacker and Karunagaran 2015). Preclinical *in vitro* and *in vivo* data have shown that CUR is one of the effective treatments for brain tumors including glioblastoma multiforme. The anticancer effects are said to potentiate by CUR ability to induce G2/M cell cycle arrest, activate apoptotic pathways, induct autophagy, disrupt molecular signaling, inhibit invasion and metastasis, and increase the efficacy of existing chemotherapeutics (Klinger and Mittal 2016). Ironically, in a recent study, CUR has been reported to hinder the antitumor effect of standard anticancer drugs (i.e., vinblastine) via the inhibition of microtubule dynamics and MMP in HeLa cervical cancer cells. Vinblastine-induced microtubule depolymerization and cell death were reduced in HeLa cells pretreated with CUR compared to the control, and CUR also reduced ROS production by vinblastine. In light of the above finding, it is confirmed that patients treated with vinblastine should not consume CUR (Lee et al. 2016a). This compound should be watched for interactions when administered as adjuvant therapy. Further studies are warranted on these lines.

Several anticancer mechanisms of CUR have been discussed for more than two decades. To date, most studies show that CUR can be a better drug candidate to induce apoptosis-mediated cell death and cell cycle arrest in several cancer cell lines. It is worth mentioning that this compound interferes with various signaling pathways, such as TGF-β, Wnt/β-catenin, etc. Further, it was reported that CUR is one of the modulators of P-glycoprotein (P-gp). Overexpression of P-gp is said to be one of the main mechanisms involved in the development of multidrug resistance. P-gp (encoded by the MDR1 gene, also referred to as ABCB1) is a drug-efflux pump from the ATP-binding cassette transporters family, which efficiently removes cytotoxic drugs from the intracellular environment through an ATP-dependent mechanism (Lopes-Rodrigues et al. 2016). In view of the previously mentioned study, it is clear that during combination therapy, CUR can prolong the bioavailability of standard anticancer drugs by inhibiting the P-gp that are responsible for the removal of cytotoxic drugs.

In order to translate the preclinical antitumor effects of CUR into clinical practice, few clinical trials have been performed so far. Currently, CUR is in phase II clinical trial for prostate cancer (Mahammedi et al. 2016). Several phase II clinical trials on the antitumor effects of CUR in pancreatic cancers were conducted. The first clinical trial was conducted in 2008 in pancreatic cancer patients and studied the efficacy of CUR used as a monotherapy in 25 pancreatic cancer patients. This study reported that there was tumor regression (73%), a significant increase in serum cytokine levels (IL-6, IL-8, IL-10, and IL-1 receptor antagonists), and downregulation of the expressions of nuclear factor-kappa B (NF-κB), cyclooxygenase-2 (COX-2). Moreover, no significant toxicity was observed in patients who received CUR (Dhillon et al. 2008). In another study, a phase I/II clinical trial was conducted with CUR in 21 patients with pancreatic cancer (resistant to gemcitabine-based chemotherapy), combining gemcitabine-based chemotherapy with CUR treatment (8 g daily oral dose). Results from this study indicate that CUR daily with gemcitabine-based chemotherapy was safe and feasible in patients with pancreatic cancer. The daily oral dose of CUR of 8 g or less is the most commonly used in clinical trials, due to its poor bioavailability. Another interesting synergistic study has tested the efficacy and feasibility of CUR (8 g daily oral dose) in combination with gemcitabine monotherapy in 17 chemo-naïve patients with pancreatic cancer. In contrast to previous studies, increased gastrointestinal toxicity was observed in seven patients treated with combination therapy, and it was reported that toxicity was probably due to the elevated dose of CUR combined with gemcitabine. For this reason, the dose of CUR was reduced from 8 to 4 g and used clinically later on (Epelbaum et al. 2010).

From several clinical studies, the effects of CUR on breast cancer were reported. For instance, in a phase IIa clinical trial of CUR for the prevention of colorectal neoplasia, CUR was well tolerated at both 2 g and 4 g in patients and could decrease the aberrant crypt foci (ACF) number (Carroll et al. 2011). In addition, CUR treatment improved patients with CRC via upregulation of p53 expression in tumor cells and consequently speeded up tumor cell apoptosis (He et al. 2011).

As of 2014, over 65 clinical trials conducted on molecules obtained from the plant *Curcuma longa* and have highlighted the protective role of CUR in various chronic diseases, including autoimmune, cardiovascular, neurological, and psychological diseases, diabetes, and cancer (Prasad et al. 2014). As of 2015, 116 studies regarding the diverse actions of CUR could be found, and among these 99 studies were based on the anti-inflammatory properties of CUR. As of November 2016, 137 clinical trials were conducted (http://www.clinicaltrials.gov/). The most noticeable diseases for which trials had been conducted were cancer (e.g., lung, prostate, breast, pancreatic, and colorectal), rheumatoid arthritis, and inflammatory bowel diseases (IBD—ulcerative colitis and Crohn's disease), which reflect the pleiotropic actions of curcumin (Table 9.1). Despite its effective anticarcinogenesis properties, CUR has poor solubility, instability, and extensive metabolism result in poor oral bioavailability, and it is one of the major hurdles to achieving an effective biological concentration in an affected area during cancer chemotherapy.

Epigallocatechin-3-gallate (EGCG)

Epigallocatechin-3-gallate (EGCG), one of the major natural catechins found in green tea (*Camellia sinensis*), has the potential impact on a variety of human diseases. The chemopreventive potentials of green tea extract have been demonstrated in animal on different organ models of cancer, such as skin, lung, oral cavity, esophagus, forestomach, stomach, small intestine, colon, pancreas, and mammary gland (Yang et al. 2002). The major catechins in green tea are (–)-epicatechin-3-gallate, (–)-epigallocatechin, and (–)-epicatechin (Figure 9.3). EGCG is the most abundant in green tea and accounts for 50%–80% representing 200–300 mg/brewed cup of green tea (Khan et al. 2006).

The cytotoxic effects of the green tea-derived phytochemicals have been studied against HCT-116 and SW-480 human CRC cells, and it seems that among the 10 polyphenols (caffeic acid, gallic acid, catechin, epicatechin, gallocatechin, catechin gallate, gallocatechin gallate, epicatechin gallate, EGCG), EGCG had the most potent antiproliferative effects, and was demonstrated to induce significant cell cycle arrest at G1 phase and apoptosis (Du et al. 2012). The cancer-preventive effects of EGCG are widely supported by results from epidemiological, cell culture, animal, and clinical studies. EGCG is a known antioxidant compound, and it is proposed that this flavonoid acts against inflammation, proliferation, and initiation of carcinogenesis. Therefore, here we discuss in detail the anticancer potentials of EGCG.

Epigallocatechin gallate has been identified as a powerful antioxidant, preventing oxidative damage *in vivo*, but also as an anti-angiogenic and antitumor agent and as a modulator of tumor cell response to chemotherapy. The anticancer properties of EGCG present in green tea extract are said to induce apoptosis and cell cycle arrest by altering the expression of cell cycle regulatory proteins, activating caspases, and suppressing oncogenic transcription factors. *In vitro* studies have demonstrated that EGCG blocks carcinogenesis by affecting a wide array of signal transduction

TABLE 9.1

Human Studies Using Plant-Derived Compounds against Different Types of Cancer

Title	Identifier Number	Medical Condition	Phase of Clinical Trial	Status
		Curcumin		
Curcumin for the prevention of radiation-induced dermatitis in breast cancer patients	NCT01042938	Breast cancer	Phase 2	Completed
Effect of curcumin addition to standard treatment on tumor-induced inflammation in endometrial carcinoma	NCT02017353	Endometrial carcinoma	Phase 2	Completed
Trial of curcumin in advanced pancreatic cancer	NCT00094445	Pancreatic neoplasms adenocarcinoma	Phase 2	Completed
Curcumin biomarker trial in head and neck cancer	NCT01160302	Head and neck cancer	Phase 0	Completed
Curcumin for the prevention of colon cancer	NCT00027495	Colorectal cancer	Phase 1	Completed
Oral curcumin for radiation dermatitis	NCT01246973	Radiation-induced dermatitis	Phase 2 Phase 3	Completed
An open-label prospective cohort trial of curcumin plus tyrosine kinase inhibitors for EGFR-mutant advanced NSCLC	NCT02321293	Lung cancer	Phase 1	Ongoing
		Curcumin with Standard Anticancer Drugs		
Gemcitabine with curcumin for pancreatic cancer	NCT00192842	Pancreatic cancer	Phase 2	Completed
Curcumin in combination with 5FU for colon cancer	NCT02724202	Metastatic colon cancer	Phase 0	Ongoing
		Green Tea–derived Polyphenols		
Green tea catechins in treating patients with prostate cancer undergoing surgery to remove the prostate	NCT00459407	Prostate cancer	Phase 1	Completed
Green tea intake for the maintenance of complete remission in women with advanced ovarian carcinoma	NCT00721890	Ovarian carcinoma	Phase 2	Completed
Green tea catechin extract in preventing esophageal cancer in patients with Barrett's esophagus	NCT00233935	Esophageal cancer	Phase 1	Completed

(Continued)

TABLE 9.1 (*Continued*)

Human Studies Using Plant-Derived Compounds against Different Types of Cancer

Title	Identifier Number	Medical Condition	Phase of Clinical Trial	Status
Green tea extract in preventing cancer in former and current heavy smokers with abnormal sputum	NCT00573885	Lung cancer tobacco use disorder	Phase 2	Completed
Green tea and reduction of breast cancer risk	NCT00917735	Breast cancer	Phase 2	Completed
Polyphenon E in treating patients with high risk of colorectal cancer	NCT01606124	Colorectal cancer	Phase 2	Completed
Green tea extract in preventing cervical cancer in patients with human papillomavirus and low-grade cervical intraepithelial neoplasia	NCT00303823	Cervical cancer	Phase 2	Completed
Topical green tea ointment in treatment of superficial skin cancer	NCT02029352	Carcinoma, basal cell	Phase 2 Phase 3	Completed
Oral green tea extract for small cell lung cancer	NCT01317953	Small cell lung carcinoma	Phase 1	Ongoing
Green tea extract in treating patients with nonmetastatic bladder cancer	NCT00666562	Bladder cancer	Phase 2	Ongoing
Resveratrol				
Resveratrol for patients with colon cancer	NCT00256334	Colon cancer	Phase I	Completed
A clinical study to assess the safety, pharmacokinetics, and pharmacodynamics of SRT501 in subjects with colorectal cancer and hepatic metastases	NCT00920803	Neoplasms, colorectal	Phase 1	Completed
Resveratrol in treating patients with colorectal cancer that can be removed by surgery	NCT00433576	Adenocarcinoma of the colon and rectum	Phase 1	Completed
Prostate phytochemical and PUFA intervention	(EudraCT Number: 2006-006679-18)	Localized prostate cancer	NS*	Ongoing
A biological study of resveratrol's effects on Notch-1 signaling in subjects with low grade gastrointestinal tumors	NCT01476592	Neuroendocrine tumor	*	Ongoing

(*Continued*)

TABLE 9.1 (*Continued*)

Human Studies Using Plant-Derived Compounds against Different Types of Cancer

Title	Identifier Number	Medical Condition	Phase of Clinical Trial	Status
UMCC 2003-064 resveratrol in preventing cancer in healthy participants (IRB 2004-535)	NCT00098969	Unspecified adult solid tumor	Phase 1	Completed
Silymarin/silibinin				
Siliphos in advanced hepatocellular carcinoma	NCT01129570	Advanced hepatocellular carcinoma	Phase 1	Completed
Silymarin (milk thistle extract) in treating patients with acute lymphoblastic leukemia who are receiving chemotherapy	NCT00055718	Drug/agent toxicity by tissue/organ Leukemia	Phase 2	Completed
The effect of high-dose silybin-phytosome in men with prostate cancer	NCT00487721	Prostate cancer	Phase 2	Completed
Evaluation of effects of silymarin on cisplatin-induced nephrotoxicity in upper gastrointestinal adenocarcinoma	NCT01829178	Upper GI cancer	Phase 2 Phase 3	Completed
Estrogen receptor beta agonists (Eviendep) and polyp recurrence (CRC)	NCT01402648	Adenocarcinoma of colon Recurrent	Phase 1 Phase 2	Completed
Genistein				
Genistein in preventing breast or endometrial cancer in healthy postmenopausal women	NCT00099008	Breast cancer Endometrial cancer	Phase 1	Completed
Phase II study of isoflavone G-2535 (genistein) in patients with bladder cancer	NCT00118040	Recurrent bladder cancer Stage I, II, III	Phase 2	Completed
Effects of genistein combined polysaccharide on prostate cancer	NCT00269555	Prostate cancer	*	Completed
Genistein and interleukin-2 in treating patients with metastatic melanoma or kidney cancer	NCT00276835	Kidney cancer Melanoma (skin)	Phase 0	Completed
Genistein in treatment of metastatic colorectal cancer	NCT01985763	Colon, rectal and colorectal cancer	Phase 1 Phase 2	Ongoing
Genistein with Standard Anticancer Drugs				
Gemcitabine hydrochloride and genistein in treating women with stage IV breast cancer	NCT00244933	Breast cancer	Phase 2	Completed

(*Continued*)

TABLE 9.1 (*Continued*)

Human Studies Using Plant-Derived Compounds against Different Types of Cancer

Title	Identifier Number	Medical Condition	Phase of Clinical Trial	Status
MTD determination, safety and efficacy of the decitabine–genistein drug combination in advanced solid tumors and non-small cell lung cancer	NCT01628471	Non-small cell lung cancer	Phase 1 Phase 2	Completed
Cholecalciferol and genistein before surgery in treating patients with early stage prostate cancer	NCT01325311	Prostate adenocarcinoma stage I, IIA, IIB prostate cancer	Phase 2	Completed
Genistein, gemcitabine, and erlotinib in treating patients with locally advanced or metastatic pancreatic cancer	NCT00376948	Pancreatic cancer	Phase 2	Completed

Note: For further details, see U.S. Clinical Trial Registry (https://clinicaltrials.gov/) and European Clinical Trial Registry (https://www.clinicaltrialsregister.eu/).

* Not specified.

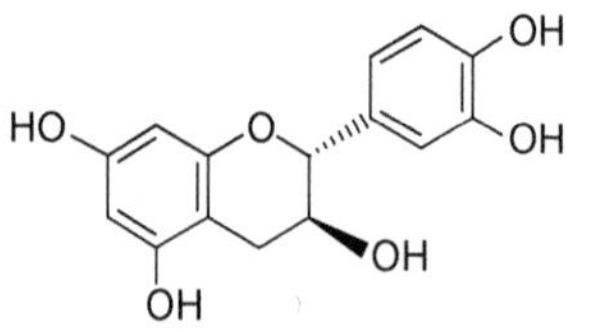

Catechin
($C_{15}H_{14}O_6$; CAS. No. 18829-70-4)

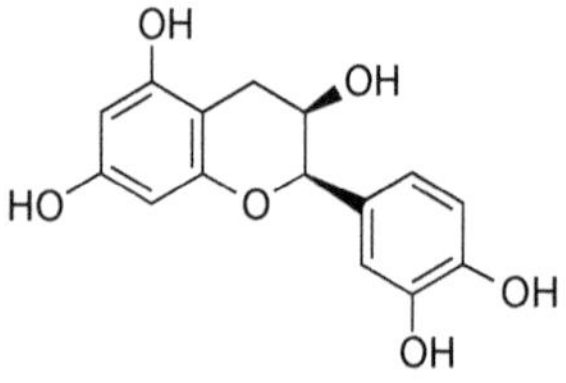

(–)-Epicatechin
($C_{15}H_{14}O_6$; CAS. No. 490-46-0)

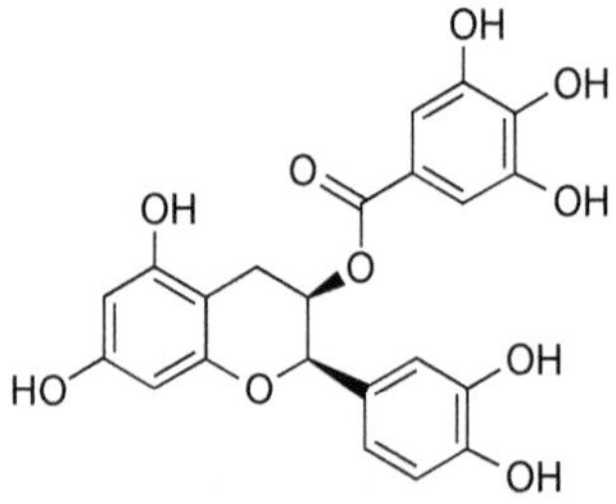

(–)-Epicatechin gallate
($C_{22}H_{18}O_{11}$; CAS. No. 989-51-5)

FIGURE 9.3 Chemical structure of major polyphenols in green tea extract.

pathways including JAK/STAT, mitogen-activated protein kinase (MAPK), PI3 K/ AKT (phosphoinositide-3-kinase/protein kinase B), Wnt, and Notch. EGCG stimulates telomere fragmentation through inhibiting telomerase activity; further, several clinical studies have revealed that treatment by EGCG inhibits solid tumor incidence in different organs, such as liver, stomach, skin, lung, mammary gland, and colon (Singh et al. 2011).

Epigenetic mechanisms have been implicated in the chemoprevention activity of EGCG, especially against skin cancer (Nandakumar et al. 2011). Epigenetic alterations, in particular, aberrant DNA methylation and acetylation of nonhistone proteins associated with inappropriate gene silencing, contribute significantly to the initiation and progression of human cancer. Reactivation of some methylation-silenced genes by EGCG was demonstrated in human colon cancer HT-29 cells, esophageal cancer KYSE 150 cells, and prostate cancer PC3 cells (Fang et al. 2003). In skin cancer, EGCG inhibits cancer-associated stages and exhibits an inhibitory effect on DNA methylation via blocking performance of DNA methyltransferases (DNMTs) (Nandakumar et al. 2011; Fang et al. 2007).

Autophagy is one of the arenas that was implicated in recent anticancer strategies. It is the process by which cellular material is delivered to lysosomes for degradation and recycling. It is involved in cell growth, survival, development, and cell death. Autophagy plays dual roles in the formation and progression of cancer, including both suppressive and promotive roles. According to Thorburn et al., the roles of autophagy in cancer treatment are complicated by two important discoveries over the past few years. First, most (perhaps all) anticancer drugs, as well as ionizing radiation, affect autophagy. In most, but not all cases, these treatments increase autophagy in tumor cells. Second, autophagy affects the ability of tumor cells to die after drug treatment, but the effect of autophagy may be to promote or inhibit cell death (Thorburn et al. 2014). In an oral cancer model, EGCG treatment has been found to suppress proliferation of SSC-4 human oral squamous cell carcinoma (OSCC) cells in a dose- and time-dependent manner. Concomitantly, the activation of apoptosis and autophagy in response to EGCG exposure in SSC-4 cells was observed. Treatment with EGCG activates the cell death receptors, leading to activation of intrinsic apoptotic pathways. All of these results suggest that EGCG has excellent potential to become a therapeutic compound for patients with OSCC by inducing tumor cell death via apoptosis and autophagy (Irimie et al. 2015).

Epidemiological evidence indicates the association between tea consumption and decreased risk of breast cancer, and this was also confirmed by population-based case-control studies carried out in Asian and U.S. populations. There were studies showing that green tea consumption significantly reduced the risk of breast cancer (Li et al. 2016; Kumar et al. 2009; Inoue et al. 2008). It was shown that short-term exposure of breast cancer cells to 4-(methylnitrosamino)-1-(3-pyridyl)-1-butanone and benzo[a]pyrene would increase the level of ROS, resulting in activation of the extracellular signal-regulated kinase (ERK) pathway and subsequent induction of DNA damage. *In vitro* and *in vivo* studies showed that tea catechins prevented breast carcinogenesis by alleviating ROS stress (Rathore et al. 2012; Kaur et al. 2007; Ruch et al. 1989). Several anticancer mechanisms of green tea polyphenols have been discussed. However, the anticarcinogenic activity is considered to be related to their protection of DNA from ROS-induced damages by alleviating ROS stress. The PI3 K/Akt/mTOR (mammalian target of rapamycin) signaling pathway is a commonly activated signaling pathway in human cancer. Apparently, EGCG was confirmed to be an ATP-competitive inhibitor of both PI3 K and mTOR in breast cancer cells (MDA-MB-231), and molecular docking studies also confirmed that EGCG binds well to the PI3 K domain-active site, thereby showing ATP-competitive activity (Oliveira et al. 2016; Van Aller et al. 2011). In the breast cancer cell line T47D, catechin phosphorylated JNK/stress-activated protein kinases (SAPK) and p38, and this in turn inhibited the phosphorylation of cdc2, and regulated the expression of cyclin A, cyclin B1, and cyclin-dependent kinases (CDKs) proteins, thereby causing G2 arrest (Deguchi et al. 2002).

The green tea phytochemicals have been widely studied for their beneficial effect against various gastrointestinal-related cancers. For example, the cytotoxic effect of EGCG-induced cell death in HT-29 cells has been reported. EGCG treatment to colon cancer cells resulted in a strong activation of AMPK (AMP-activated protein kinase) and an inhibition of COX-2 expression. The activation of AMPK was accompanied by the reduction of VEGF (vascular endothelial growth factor) and glucose transporter (Glut-1) in EGCG-treated cancer cells (Hwang et al. 2007).

As we have discussed, this section summarizes recent research focusing on EGCG-induced cellular signal transduction events that seem to have implications in the inhibition of cell proliferation and transformation, induction of apoptosis, and autophagy of preneoplastic and neoplastic cells, as well as inhibition of angiogenesis, tumor invasion, and metastasis. Based on the epidemiological evidence, it is obvious that reduction in breast cancer risk was often associated with green tea consumption, rather than black tea consumption, because catechins, especially EGCG, are the major bioactive components in green tea appraised for their anticancer activity. Black tea is said to be fully fermented tea, and about 80% of tea catechins are oxidized and converted into orange and red tea pigments (theaflavins and thearubigins) during fermentation.

RESVERATROL

Resveratrol (trans-3,4′,5-trihydroxystilbene) (Figure 9.4) is a pleiotropic polyphenol belonging to the stilbene family. Resveratrol (RES) is found in an abundant amount in red wine, grape berry skins and seeds, and especially in the dried roots of the plant *Polygonum cuspidatum*. Resveratrol content in grapes varies from 0.16

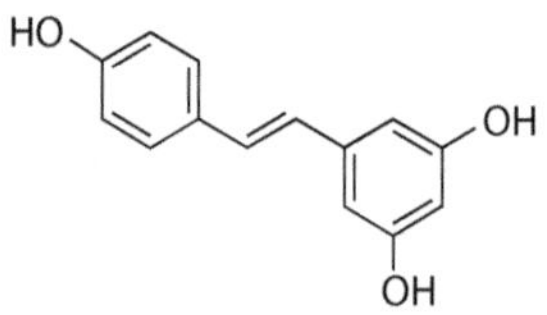

FIGURE 9.4 Chemical structure of resveratrol ($C_{14}H_{12}O_3$; CAS No. 501-36-0).

to 3.54 mg/g; dry grape skin contains about 24 mg/g of RES. It is also present in other berries and nuts, for instance, cranberry raw juice contains about 0.2 mg/L. In other natural foods, the concentration of RES varies in the range of mg/g (peanuts, pistachios) to ng/g (bilberries, blueberries). It has been documented that red wine contains a much greater number of polyphenolic compounds than white wine. The concentration of RES ranges from 0.1 to 14.3 mg/L in various types of red wine, while white wine contains only about 0.1–2.1 mg/L of RES (Hu et al. 2013; Mukherjee et al. 2010).

Current scientific evidence indicates that at lower doses RES acts as an anti-apoptotic and cardioprotective agent. At the same time, at higher doses it exhibits pro-apoptotic potentials in cancer cells (Vanamala et al. 2010; Hung et al. 2000). It is worth mentioning that RES affects various intracellular mediators, participating in all three stages of oncogenesis: initiation, promotion, and progression. Depending on a tumor model, RES intracellular targets can be nitric oxide (NO), p53, apoptosis regulators, COXs, transcription factors, cyclins, calpains, caspases, interleukins, cathepsins, etc. (Aggarwal et al. 2004).

Sirtuins are NAD (+)-dependent histone deacetylases regulating important metabolic pathways in prokaryotes and eukaryotes and are involved in many biological processes such as cell survival, senescence, proliferation, apoptosis, DNA repair, and cell metabolism. The seven members (silent information regulators 1–7) of this family of enzymes are considered potential targets for the treatment of human pathologies including cancer (Carafa et al. 2016). Resveratrol induces apoptosis through several different pathways: receptor-mediated or caspase-8-dependent pathway; mitochondrial or caspase 9–dependent pathway or cell cycle arrest, and the pathway affecting SIRT 1 (Han et al. 2015).

Several research groups have investigated the effects of RES on breast cancer cells *in vitro* and *in vivo*. However, recent evidence indicates that RES may inhibit cancer progression through modulation of microRNAs (miRNAs). Resveratrol appears to regulate apoptotic and cell cycle machinery in breast cancer cells by modulating key tumor-suppressive miRNAs including miR-125b-5p, miR-200c-3p, miR-409-3p, miR-122-5p, and miR-542-3p. In another study, RES-mediated miRNA modulation has been said to regulate key anti-apoptotic and cell cycle proteins including bcl-2, and CDKs, which are critical for its activity (Venkatadri et al. 2016). Resveratrol inhibited breast cancer cell proliferation by stimulating SIRT-1. Activation of the AMPK pathway leads to mTOR activation, which stimulates the cell proliferation. It was observed that RES can block AMPK phosphorylation (overexpressed in tumor cells) by SIRT-1 activity (Lin et al. 2010). In a study using breast cancer cells, RES has been reported to upregulate the pro-apoptotic gene expression (BAD, p53, TP53I3, p21, c-fos, and COX-2) and significantly reduced proliferation (Chin et al. 2014). In synergistic studies, RES appears to improve the anticancer effects of doxorubicin in combination, through inhibiting breast cancer cell proliferation and invasion, and inducing apoptosis via suppression of chronic inflammation and autophagy (Rai et al. 2016).

The synthetic analogue of resveratrol, HS-1793, was investigated on the proliferation and apoptotic cell death using HCT116 human colon cancer cells. In this study, HS-1793 induced apoptosis through alteration of the bax/bcl-2 expression ratio, and caspase activations. This study also shows HS-1793 induced G2/M arrest in the cell cycle progression in HCT116 cells (Kim et al. 2016). In other studies, RES was found to induce ROS generation and mitochondrial respiration and apoptosis in SW620 colon cancer cells (Blanquer-Rosselló et al. 2017). Moreover, the suppression of Wnt/β-catenin signaling and elevated mitochondrial-mediated apoptosis and inhibition of proliferation, sphere formation, and nuclear translocation of β-catenin (a critical regulator of CCC proliferation) were also reported upon RES treatment in colon CCCs *in vitro*. In addition, RES along with grape extract seed treatment has reduced the number of crypts containing cells with nuclear β-catenin via induction of apoptosis *in vivo* (Reddivari et al. 2016). These findings suggest that RES have potential as a candidate chemotherapeutic agent against human colon cancer.

The effects of RES on cell proliferation of hepatocarcinoma cells with high concentration of glucose were evaluated in another study. Interestingly, results from this study showed that high glucose concentration upregulated activated STAT-3 and enhanced cellular viability. Resveratrol was able to suppress proliferation, and high glucose concentration induced activation of STAT-3 and Akt in hepatocarcinoma cells (Li et al. 2013).

The mammalian or mechanistic target of rapamycin complex 1 (mTORC1) is one of the signaling pathways and is hyperactivated in a variety of cancers and tumor syndromes; hence, mTORC1 inhibitors (rapamycin) are being investigated for the treatment of various cancers. Resveratrol has been targeted for mTORC1 cancer signaling. The combination of rapamycin and RES selectively promoted apoptosis of cancerous cells with mTOR pathway hyperactivation. Moreover, this combination prevented tumor growth and lung metastasis when tested in mouse models (Alayev et al. 2015).

An *in vivo* perspective was taken by Jang and colleagues who were the first to show that RES may act as a chemopreventive agent when they found that topical application of the compound was able to inhibit tumor formation in the two-stage skin cancer model in mice (Jang et al. 1997). Studies then concretely proved that in mouse models of skin tumorigenesis, topical resveratrol prevented tumor formation through promoting apoptosis, G2/M phase cell cycle arrest, and inhibiting COX activity and PG production (Hu et al. 2016; Tsai et al. 2012). Resveratrol treatment (50 mg/kg) reduced tumor multiplicity in a dose-dependent manner in DEN-treated cancerous animals, after 14 weeks of tumor initiation (Luther et al. 2011). In a recent systematic review, it was reported that RES treatment causes a significant reduction in tumor incidence when compared with the control groups *in vivo* as well as *in vitro* studies (Feng et al. 2016).

In summary, the studies we discussed in this section show that RES supplementation could potentially have many positive health benefits including decreased cancer risk. Regarding its molecular role, it appears that RES, at least in high concentrations as used in most studies, affects most signaling and metabolic pathways. This is a shift in the insight that RES directly regulates SIRT 1 to the current focus on cAMP signaling. Resveratrol should be watched for its effects on SIRT 1. However, there are limited clinical trials with small sample sizes, and animal models have had mixed results. Hence, there is a need for more extensive and consistent studies in animal models.

SILYMARIN AND ITS ACTIVE CONSTITUENTS

Silybum marianum and its seeds, one of the most commonly consumed dietary supplements and it is used clinically as an antihepatotoxic agent worldwide including
in the United States, contain natural compounds, called *flavonolignans*. Silymarin
(SIL) is a dry mixture of these compounds, which are extracted after processing
with ethanol, methanol, and acetone. Silymarin contains mainly silybin A, silybin B,
taxifolin, isosilybin A, isosilybin B, silichristin A, silidianin, and other compounds
in smaller concentrations (Tyagi et al. 2002) (Figure 9.5).

The terms *milk thistle, silymarin,* and *silybin* are generally used interchangeably; however, each of these has specific characteristics and actions, with an intrinsic
beneficial or toxic effect. These compounds have been studied extensively for their
beneficial effect against human prostate adenocarcinoma cells, estrogen-dependent
and estrogen-independent breast carcinoma cells, cervical, hepatoma cancer cells,
colon cancer cells, and NSCLC and have come out with promising results (Mateen
et al. 2010; García-Maceira and Mateo 2009; Tyagi et al. 2002). Silibinin (SBN)
has been reported to inhibit growth in a variety of cancer cells (Tyagi et al. 2002).
Silibinin has been shown to inhibit multiple cancer cell signaling pathways in preclinical models, demonstrating promising anticancer effects *in vitro* and *in vivo*. It
has been well studied for its action against prostate cancer (PC) (Zi et al. 1998). The
phytochemical constituents of the milk thistle plant have been tested against PC
in vitro. This study reports that isosilybin B is one of the active ingredients among
milk thistle phytochemicals (Deep et al. 2008). The anticancer efficacy of SIL in
human prostate carcinoma DU145 cells was previously shown. Since then several
studies have reported the anticancer effect of these flavonolignans (Zi and Agarwal

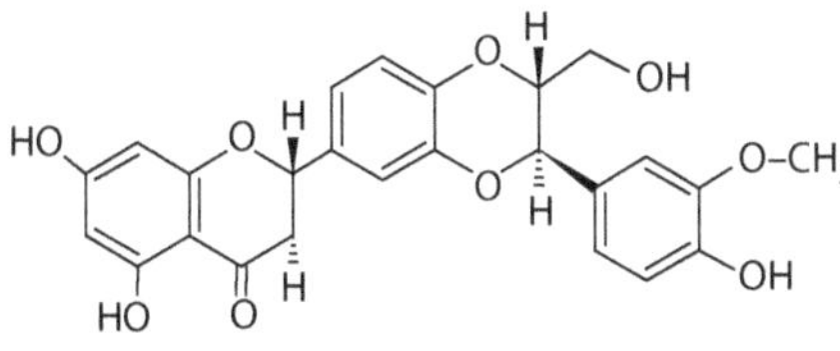

Silybin/Silibinin
$(C_{25}H_{22}O_{10};$ CAS. No. 22888-70-6)

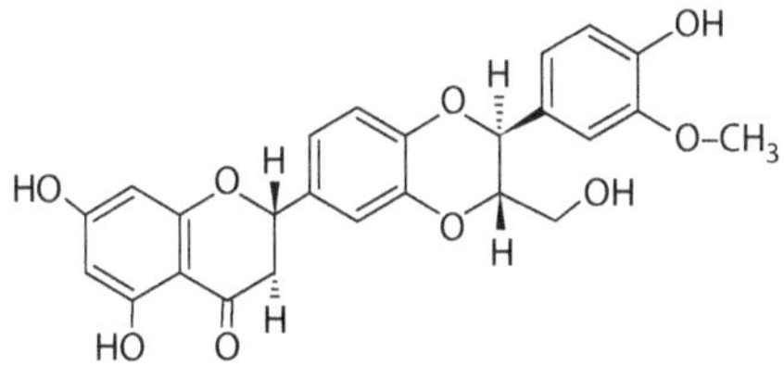

Isosilibinin
(CAS. No. 142796-21-2)

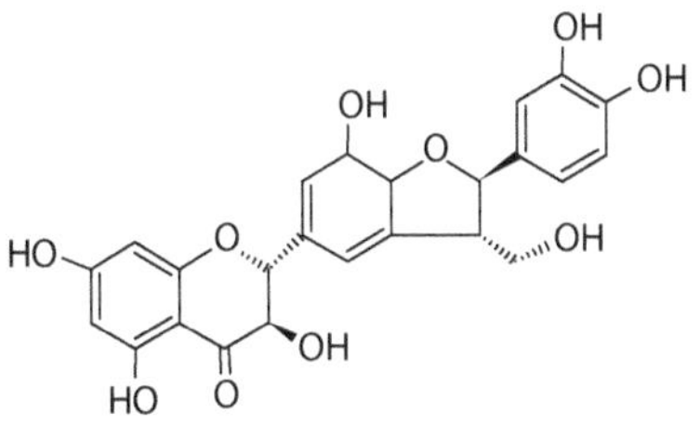

Silychristin
(CAS. No. 33889-69-9)

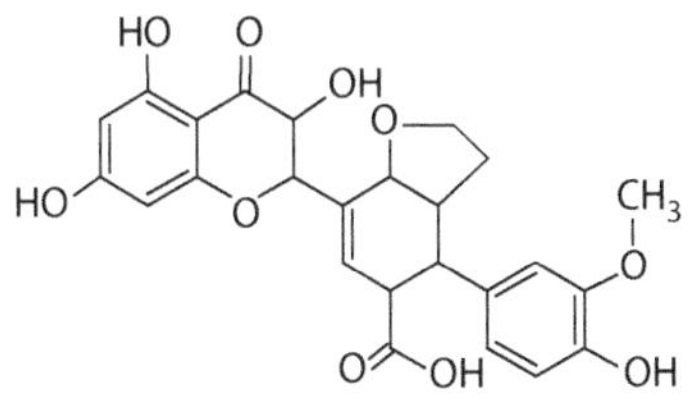

Silydianin
(CAS. No. 29782-68-1)

FIGURE 9.5   Active constituents of SIL.

1999). The mechanism of anticarcinogenic effect of SIL treatment on PC cells is likely through impairment of the epidermal growth factor receptor-1 (ErbB-1)-SHC-mediated signaling pathway, induction of CDKIs, and G1 phase of cell cycle arrest (Zi et al. 1998), and this erbB-1 and other members of the erbB family have been shown to play imperative roles in human PC.

In combination studies, SBN strongly synergized the growth-inhibitory effect of doxorubicin (a known anticancer drug used in prostate carcinoma) in DU145 cells through G2/M cell cycle arrest. Interestingly, this combination (SBN with doxorubicin) caused 41% apoptotic cell death compared with 15% by either agent alone. In addition, a SBN and doxorubicin combination was shown to effectively inhibit the growth of androgen-dependent prostate carcinoma LNCaP cells (Tyagi et al. 2002). In other studies, it was reported that SBN has the ability to induce apoptosis of human PC cell lines such as PC-3, LNCaP, and RWPE-1 through autophagy, generation of ROS, and activation of ER stress that requires the disruption of Ca^{2+} homeostasis through ROS generation. Those ROS were produced, in part, from the mitochondrial NADPH oxidase system, upstream pathway of Ca^{2+} signaling (Kim et al. 2016). In light of the above studies, it is clear that SBN either alone or in combination with other drugs can exhibit profound anticancer effects, especially against PC. Hence, it is suggested that this compound can be used as adjuvant therapy along with standard chemotherapeutic agents. However, further animal and clinical studies are warranted on these lines.

Studies from us and others have concretely reported the antihepatotoxic potentials of SBN in experimental animals against various drugs and chemical-induced hepatotoxicity (Ezhilarasan et al. 2016; Ezhilarasan and Karthikeyan 2016; Raghu et al. 2015; Liang et al. 2015; Ezhilarasan et al. 2012). In a previous study, SBN was tested against human cervical (HeLa) and hepatoma (Hep3B) cells. From these studies, the anticancer activities of SBN have been reported via inhibition of PI3 K/Akt, and it was associated with cell survival and angiogenesis, which are essential for the adaptation of cancer cells to microenvironmental hypoxia and hence for tumor progression (García-Maceira and Mateo 2009).

Silibinin has been found to possess an inhibitory effect against MDA-MB-231, MCF7 breast cancer cells. It is reported that SBN inhibits the proliferation and induces apoptosis of MCF-7 cells by downregulating BAK1, p53, p21, BRCA1, BCL-Xl (Bayram et al. 2017; Pirouzpanah et al. 2015). The STAT3 is constitutively activated in many different types of cancer and plays a pivotal role in tumor growth and metastasis. Retrospective studies have established that STAT3 expression or phospho-STAT3 (pSTAT3 or activated STAT3) are prognostic markers for breast, colon, prostate, and NSCLCs. For this reason, studies have targeted the pSTAT3 and such studies report that SBN inhibits pSTAT3 expression in preclinical studies of prostate, breast, skin, gastric, and lung cancers (Bosch-Barrera and Menendez 2015). Colorectal cancer is still the third most common cancer in the world (Mohammadi et al. 2016). Silibinin exhibits cytotoxic effects and pro-apoptotic potential in human CRC COLO 205 and HT-29 cells (Tsai et al. 2015; Akhtar et al. 2014). In cervical carcinoma cells, SBN and SIL treatments inhibit cell growth and DNA synthesis (Bhatia et al. 1999). Silibinin treatment to HeLa cells resulted in a G2 phase of cell cycle arrest and induced a decrease in CDKs involved in both G1 and G2 cell cycle progression. Silibinin was shown to induce both dose- and time-dependent apoptotic death in HeLa cells via the intrinsic and extrinsic apoptotic pathways (Zhang et al. 2012).

In clinical translational studies, the compounds obtained from the medicinal plant milk thistle have also been studied for their effects against various other cancers in human subjects. Silipide (a SBN formulation) was given orally to patients with confirmed colorectal adenocarcinoma at doses 360–1440 mg of SBN per day for 7 weeks. It was reported that intervention with SBN was found to be ineffective in modulating the circulating levels of insulin-like growth factor binding protein 3 (IGFBP-3), IGF-1 and pyrimidopurinone adduct of dexoguanosine, a marker for oxidative DNA damage; but it is worthy to note that high levels of SBN were achieved in colorectal mucosa of the patients, which further supports the need to conduct more elaborate clinical trials to evaluate its efficacy as a potential chemopreventive agent (Hoh et al. 2006).

The silybin-phosphatidylcholine (Siliphos) combination is an orally bioavailable complex of SBN that has been previously reported for its hepatoprotective effects in experimental animals (Jesudas et al. 2016). This compound has been known for higher bioavailability than conventional preparations with organic solvents such as propylene glycol. To achieve effective doses and bioavailability in human subjects, silybin-phytosome was administered orally to prostate cancer patients at a dose range from 2.5 to 20 g daily, in three divided doses for 4 weeks' duration of each course. It was reported that about 13 g of oral silybin-phytosome daily, in three divided doses, appears to be well tolerated in patients with advanced prostate cancer (Flaig et al. 2007). This study provided the best evidence that SBN can be administered to humans at doses producing anti-cancer-relevant concentrations, with minimal or no side effects. It was reported later by the same investigators that high-dose oral silybin-phytosome achieves high blood concentrations rapidly, but low levels of SBN are seen in prostate tissue, and this could be due to lack of tissue penetration and may be explained by its short half-life or an active process removing SBN from the prostate (Flaig et al. 2010). However, a recent phase I clinical trial confirms the accumulation of silybin-phosphatidylcholine complex in breast tumor tissue after its oral administration, and it attains high blood concentrations (Lazzerroni et al. 2016). This complex also was studied in patients with advanced hepatocellular carcinoma (Siegel et al. 2014).

Overwhelming evidence suggests that SBN may function diversely and may serve as a novel therapy for cancer, such as prostatic cancer, colon cancer, breast cancer, bladder cancer, and hepatocellular carcinoma by regulating cancer growth, proliferation, apoptosis, angiogenesis, and many other mechanisms. Treatment of cancers such as prostate, breast, and cervical human carcinoma cells with SBN resulted in a highly significant inhibition of both cell growth and DNA synthesis in a time-dependent manner. The predominant role of SBN in regulating pleiotropic mitogenic signaling cascades involved in controlling prime endpoints like cell proliferation, cell survival, and cell cycle progression, has helped to define its efficacy at different stages of cancer, as well as to exploit its potential activity against several epithelial cancers including skin, prostate, colon, and lung. However, the mechanisms by which SBN exerts an anticancer effect are poorly defined. Widespread use of SIL/SBN as a hepatobiliary protective agent (Silybon) in humans and commercial availability of its formulations with augmented bioavailability further highlight the need to implement controlled clinical trials with these agents in cancer patients.

QUERCETIN

Quercetin (2-[3,4-Dihydroxyphenyl]-3,5,7-trihydroxy-4H-1-benzopyran-4-one) (QUE) is a ubiquitous flavonoid reported to have myriad health benefits. Major dietary supplements such as citrus fruits, apple, onion, broccoli, parsley, berries, green tea, and red wine comprise QUE. Quercetin (Figure 9.6), the most abundant flavonoid in Licorice extracts, has been shown to have anti-ulcer, anticancer, antioxidant, and anti-inflammatory properties.

Quercetin has been studied for its anticancer effect against colon 26 (CT26) and colon 38 (MC38) cells and was found to inhibit the cell viability of these cells and induced cell death via MAPKs mediated apoptotic pathway. Epithelial-mesenchymal transition has been shown to occur in the initiation of metastasis for cancer progression, wound healing, in organ fibrosis, and in other pathologies. Markers of EMT such as E, N-cadherin, β-catenin, and snail, were regulated by nontoxic concentrations of QUE and it is considered as one of the strategies to prevent cancer metastasis. This compound also inhibited the migration and invasion abilities of CT26 cells via expression of matrix metalloproteinases (MMPs) and tissue inhibitor of metalloproteinases (TIMPs) regulation. These *in vitro* studies were well correlated within an animal experimentation. Quercetin markedly decreased lung metastasis of CT26 cells in an experimental *in vivo* metastasis model (Kim et al. 2016). NF-κB is a signaling pathway that controls transcriptional activation of genes important for the tight regulation of many cellular processes and is aberrantly expressed in many types of cancer. Quercetin also was found to suppress the migration and invasion of human colon cancer caco-2 cells via regulating toll-like receptor 4 (TLR4)/NF-κB pathway. The expressions of metastasis-related proteins of MMP-2, MMP-9 were decreased, whereas the expression of E-cadherin was increased by QUE treatment to colon cancer cells (Han et al. 2016). In light of these studies, it is confirmed that QUE has the ability to inhibit the survival and metastatic ability of CT26 cells, and it can subsequently suppress colorectal lung metastasis in the mouse model.

Quercetin has reduced proliferation and induced cell death through mitochondrial dysfunction and significantly reduced MMP and increased production of ROS in choriocarcinoma JAR and JEG3 cell lines. Further cell cycle analysis shows a sub-G1 phase of the cell cycle arrest. In synergistic studies, the antiproliferative effect of the standard chemotherapeutic agents, cisplatin (a platinum-based drug) and

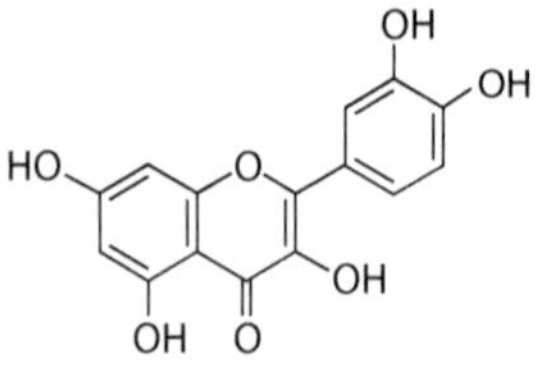

FIGURE 9.6   Chemical structure of QUE ($C_{15}H_{10}O_7$; CAS No. 117-39-5).

paclitaxel (a taxene-based drug), were enhanced when combined with QUE. This combination (cisplatin + paclitaxel + QUE) effectively inhibits the proliferation of JAR and JEG3 cells. Quercetin inhibited phosphorylation of AKT, P70S6 K, and S6 proteins, whereas it increased phosphorylation of ERK1/2, P38, JNK, and P90RSK proteins in JAR and JEG3 cells (Lim et al. 2017). The previously mentioned signaling pathways/molecules are responsible for cancer progression/suppression, and it is one of the main targets for anticancer therapies.

The anticancer effect of QUE in xenograft models with Epstein–Barr virus (EBV)-associated human gastric carcinoma was recently reported. Latent EBV infection can lead to serious malignancies, such as Burkitt's lymphoma, Hodgkin's disease, and gastric carcinoma and EBV-associated gastric carcinoma (Iizasa et al. 2012). In a xenograft mice model, QUE inhibited EBV viral protein expressions in tumor tissues from mice injected with EBV (+) human gastric carcinoma (SNU719) cells. Moreover, QUE was said to induce p53-dependent apoptosis along with the expressions of the cleaved forms of caspase-3 and 9, in EBV (+) SNU719 cells (Lee et al. 2016b).

Cell growth inhibitory effects of isomeric *in vivo* metabolites of QUE, 3-O-glucoside (active) and hyperoside (quercetin 3-O-galactoside) (inactive), myricetin, and QUE were studied previously in human bladder tumor cell lines (RT4, SCABER, and SW-780) and nontumorigenic immortalized human uroepithelial cells (SV-HUC) to compare the selectivity of the above isomeric flavonoid in cancer and noncancer cell lines. Interestingly, isorhamnetin and myricetin had very low antiproliferative effect against nontumorigenic SV-HUC cells even at very high concentrations (>200 µM) compared to bladder cancer cells, indicating the fact that their cytotoxicity is selective for cancer cells (Prasain et al. 2016).

MicroRNAs may offer novel therapeutic approaches for cancer treatment. The effect of QUE to miR profiling was performed in pancreatic ductal adenocarcinoma (PDA) cells before and after QUE treatment. It was found that miR let-7c was downregulated in pancreatic cancer. Quercetin treatment upregulated the miR let-7c expression, and this effect is implicated in its anticancer activity. Quercetin-induced miR Let-7c, in turn, decreases tumor growth by post-transcriptional activation of Numbl (Numb-like protein) and indirectly inhibits Notch signaling (Nwaeburu et al. 2016).

Head and neck squamous cell carcinoma (HNSCC) with aberrant EGFR signaling is often associated with a poor prognosis and a low survival rate. Therefore, QUE has been studied against HNSCC with aberrant EGFR signaling. Quercetin treatment effectively suppressed cell migration and invasion in EGFR-overexpressing HSC-3 and FaDu HNSCC cells. Quercetin also inhibits the colony formation of HSC-3 cells. Further, to support these findings, it has been shown to inhibit migration and invasion of the HNSCC cells via activation of gelatinase MMP-2 and MMP-9 (Chan et al. 2016).

Though QUE is appraised for its myriad anticancer properties, due to its poor water solubility as well as less bioavailability it has confined its use as a suitable anticancer drug. To counteract these obstacles, new nanoparticles formulations for drug delivery have been tried to achieve better bioavailability. One study showed that folic acid armed mesoporous silica nanoparticles (MSN-FA-Q) loaded with QUE ensure a targeted delivery with enhanced bioavailability. In breast cancer cells, MSN-FA-Q nanoformulations were shown to cause antimigratory, cell cycle arrest, and apoptosis through the regulation of Akt and Bax signaling pathways (Sarkar et al. 2016). Quercetin alone, and also in combination with RES and catechin, has been tested in breast cancer cells. Results of this study show that QUE was the most effective compound for the Akt/mTOR inhibition pathway that is a critical signaling nexus for regulating cellular metabolism, energy homeostasis, and cell growth. Treatment with QUE alone had a similar effect as the combination (QUE + RES + catechin) treatments in breast cancer cells. However, cell cycle analysis shows that QUE-treated cancer cells were arrested in the cell cycle in G2/M phase, whereas combination treatments arrested at the G1 phase of the cell cycle. *In vivo* experiments using severe combined immune deficiency mice with implanted tumors from metastatic breast cancer cells demonstrated that administration of QUE resulted in a significant reduction in tumor growth (Rivera Rivera et al. 2016).

In light of the above studies, it is confirmed that QUE mitigates proliferation and progression of cancer due to its anti-angiogenic, anti-inflammatory, and apoptotic biological effects on a variety of cancer cells. The overwhelming data on synergistic studies vigorously fortify the utilization of QUE as a chemoprevention drug, and its anti-invasive and anti-metastatic properties should be watched.

GENISTEIN

Genistein (GEN) is an isoflavonoid (estrogen-like chemical compound present in plants) that is derived from certain plant precursors by human metabolism. They are naturally occurring chemical constituents that may interact with estrogen receptors to produce weak estrogenic or anti-estrogenic effects. Soybeans contain large amounts of isoflavones, such as the GEN (Figure 9.7) and daidzein. Soybean is one of the most important food components in the Asian diet. Previously, GEN was demonstrated to be one of the predominant soy isoflavones that inhibits several steps involved in carcinogenesis. A plethora of evidence supports *in vitro* and *in vivo* anticancer effects of GEN, a soybean isoflavone (Russo et al. 2016).

This compound has been studied against TP53-depeleted human prostate carcinoma LNCaP cells. In this study, GEN treatment induced cell death in a concentration-dependent manner in cancer cells. TP53-depleted human prostate carcinoma LNCaP and NSCLC (NCI-H460) cells using shRNA targeting human TP53 were more sensitive to cell death by treatment of GEN. Genistein induced mitotic catastrophe and apoptosis via mitotic arrest by inhibiting Plk1, a conserved Ser/Thr mitotic kinase that has been identified as a promising target for anticancer drug development because its overexpression is correlated with malignancy (Shin et al. 2017). This

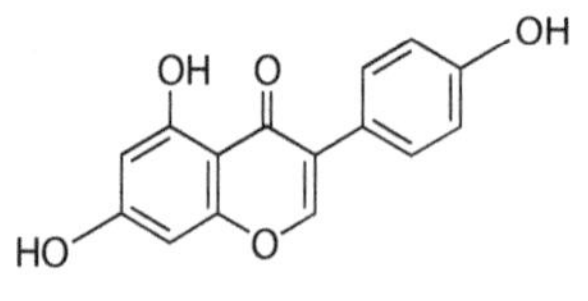

FIGURE 9.7   Chemical structure of GEN ($C_{15}H_{10}O_5$; CAS No. 446-72-0).

study suggests that GEN may be a promising anticancer drug candidate due to its inhibitory activity against Plk1 as well as EGFR and its effectiveness toward prostate and lung cancer cells.

Hepatocellular carcinoma is one of the most common malignant tumors with an accumulation of epigenetic alterations in tumor suppressor genes (TSGs), leading to hypermethylation of the genes. The hypermethylation of TSGs is associated with silencing and inactivating them. It is a well-known fact that DNA hypomethylation is the initial epigenetic abnormality recognized in human tumors. Estrogen receptor alpha (ERα) is discussed as one of the TSGs that modulates gene transcription, and its hypermethylation is because of overactivity of DNA methyltransferases (DNMTs). Taken together a recent study reported the ERα expression upon GEN treatment in HCC, and this study clearly demonstrated that GEN increases ERα expression and decreases DNMT1 expression and also inhibits proliferation and induces apoptosis in the human HCC cell line through epigenetic mechanism (Kavoosi et al. 2016). Moreover, GEN has also been tested against human HCC SNU-449 cells and has been found to inhibit the proliferation of SNU-449 cells via apoptotic cell death by DNA fragmentation and caspase activation. Proteomic analysis has shown a decrease in the thioredoxin-1 level and intracellular accumulation of ROS by GEN treatment, and it was implicated with GEN-induced apoptosis. Further, this study was able to show GEN-induced activation of apoptosis signal-regulating kinase 1, c-Jun N-terminal kinases (JNK), and p38 (Roh et al. 2016). Antiproliferative and pro-apoptotic effects of GEN have also been studied in the HCC PLC/PRF5 cell line (Dastjerdi et al. 2015).

GEN was found to be effective against 1,2-dimethyl hydrazine (DMH)-induced colon cancer in rats. In this study, it suppressed the colonic stem cell marker protein CD133, CD44, and β-catenin expressions induced by DMH (Sekar et al. 2016). In another study, GEN confers anticancer potential through inhibition of CRC cells proliferation and oral administration of GEN appears to be nontoxic to mice, and it did not inhibit tumor growth but effectively inhibited distant metastasis formation in the orthotopic implantation of human CRC tumors into mice (Xiao et al. 2015). In preclinical *in vitro* studies, GEN treatment is shown to induce apoptosis in human colon cancer LoVo, HT-29 cells by inhibiting the NF-κB pathway, as well as shows downregulation of Bcl-2 and upregulation of bax (Luo et al. 2014). GEN-inhibited cell proliferation thereby induces G2/M cell cycle arrest and apoptosis in HCT-116, with accumulation of intracellular ROS level and decrease of MMP (Wu et al. 2014), thus providing a basis for clinical application of GEN in colon cancer.

In cervical cancer studies, GEN suppressed the viability of HeLa cells in a dose-dependent manner and also caused apoptosis. Further, GEN triggered ER stress in these cells, as indicated by the upregulation of glucose-regulated protein-78 expressions (Yang et al. 2016). In another study, using HeLa cells, GEN alone or in combination with RES treatment was able to induce apoptosis by enhancing the activities of caspase-9 and caspase-3 and lowered MMP (Dhandayuthapani et al. 2013).

Epidemiological studies have revealed that high consumption of soy products is associated with low incidences of hormone-dependent cancers, including breast and prostate cancer. Genistein was observed to have structural similarity to 17β-estradiol, when compared with the other isoflavones, and to possess weak estrogenic activity.

It competes with 17β-estradiol for the estrogen receptor, with 4% binding affinity for ER-α and 87% for ER-β, thereby contributing a favorable role in the treatment of hormone-related cancers (Banerjee et al. 2008). GEN has been studied against a variety of breast cancer cell lines, and the astonishing results of these studies showed GEN induces apoptosis and promotes synergistic inhibitory effects when combined with standard anticancer drugs. For instance, GEN was shown to induce apoptosis in the low-invasive (ER-positive) MCF-7 and in the high-invasive (ER-negative) MDA-MB-231 breast cancer cell lines (Liu et al. 2005; Hsieh et al. 1998). Synergistic pro-apoptotic effects were also described when GEN was used in combination with Adriamycin and docetaxel in MDA-MB-231 cells and with tamoxifen on BT-474 breast cancer cells (Mai et al. 2007; Satoh et al. 2003). The main molecular targets of the molecule in breast cancer cells appear to be NF-κB (Li et al. 2005) and Akt pathways (Gong et al. 2003). Thus, the intake of soy products has been attributed to a lower incidence of breast and prostate cancer in Asian populations, which is mainly due to the presence of GEN. Genistein was found to inhibit angiogenesis through regulation of multiple pathways, such as regulation of VEGF, MMPs, EGFR expressions, and NF-κB, PI3-K/Akt, ERK1/2 signaling pathways, thereby causing strong anti-angiogenic effects (Varinska et al. 2015).

The *in vitro* observations have been confirmed through *in vivo* studies, suggesting that GEN exposure early in life may reduce the risk of breast cancer (Lamartiniere et al. 1998). On the contrary, in a preclinical mouse model that resulted in 17β-estradiol (E2) blood concentrations similar to those found in postmenopausal women, dietary GEN in the presence of low concentrations of circulating E2 acted in an additive manner to stimulate estrogen-dependent tumor growth *in vivo* (Ju et al. 2006). Results from this study suggest that the consumption of products containing GEN may not be safe for postmenopausal women with estrogen-dependent breast cancer. However, further studies are warranted on these lines to explore the exact role of GEN on postmenopausal women.

In synergistic studies, both GEN and topotecan (TOPO) induced cellular death in LNCaP cells, and the combination was significantly more efficacious in reducing the viability of these cells when compared with either GEN or TOPO alone. Individually or in combination with TOPO, GEN induced cell death primarily through apoptosis, through activation of caspase-3 and caspase-9, which are involved in the intrinsic pathway; and ROS levels are also found to be increased significantly with the GEN–TOPO combination treatment (Hörmann et al. 2012). This study suggests that the GEN–TOPO combination may prove to be an attractive alternative adjuvant therapy for prostate cancer.

Genistein possesses pleiotropic molecular mechanisms of action including induction of apoptosis, cell cycle arrest, ER stress and inhibition of tyrosine kinases, DNA topoisomerase II, 5α-reductase, galectin-induced G2/M arrest, protein histidine kinase, and CDK, modulation of different signaling pathways associated with the growth of cancer cells (e.g., NF-κB, Akt, MAPK, etc.). Moreover, GEN was also demonstrated as a potent inhibitor of angiogenesis. GEN is currently in phase 2 clinical trial against various pediatric oncology cases such as lymphoma, childhood lymphoma, solid tumor, childhood solid tumor, neuroblastoma, Ewing's sarcoma, Hodgkin's lymphoma, non-Hodgkin's lymphoma, rhabdomyosarcoma, soft tissue sarcoma, medulloblastoma, germ cell tumor, Wilms' tumor, brain neoplasms, child-hood medulloblastoma, and primitive neuroectodermal tumors (ClinicalTrials.gov Identifier NCT02624388).

BERBERINE

Berberine (5,6-dihydro-dibenzo [a,g] quinolizinium derivative) (Figure 9.8) is an isoquinoline quaternary alkaloid isolated from many kinds of medicinal plants such as *Hydrastis canadensis, Berberis aristata, Coptis chinensis, Coptis japon-ica, Phellondendron amurense,* and *Phellondendron chinense Schneid.* Berberine (BBR) is a major constituent of many medicinal plants of families *Papaveraceae, Berberidaceae, Fumariaceae, Menispermaceae, Ranunculaceae, Rutaceae,* and *Annonaceae* (Chen et al. 2012).

Studies have reported the pro-apoptotic potentials of BBR mediated by the impact on mitochondria. Berberine was proved to alter the MMP, inhibit mitochondrial res-piration leading to mitochondrial dysfunction, and regulate the expression of Bcl-2 family members, as Mcl-1 (Sugio et al. 2014; Kuo et al. 2005). Pro-apoptotic effects of BBR could be mediated through downregulation of the HER2/PI3 K/Akt pathway in breast cancer cells (Kuo et al. 2011) and activation of the JNK/p38 signaling path-way in human colon cancer cells (Hsu et al. 2007) and impact of BBR on the NF-κB/p65, Akt, and MAPK pathway, leading to inactivation of this factor with consequent triggering of the apoptotic process, cell cycle, and invasion pathway arrest.

Berberine has been well studied against various HCC cell lines. In such studies, BBR inactivated p38 and Erk1/2 signaling pathway through upregulation of plasmin-ogen activator inhibitor-1 (PAI-1), a tumor suppressor. This property was well cor-related with HCC cell invasion and migration activity of BBR (Wang et al. 2016a). Berberine treatment caused cytotoxicity in H22, HepG2, and Bel-7404 cells. This cytotoxic potential has directly correlated with its anticancer effects, and BBR treat-ment also reduced the weight of tumors in a H22 transplanted tumor model in mice (Li et al. 2015). This compound has also been demonstrated to induce apoptosis and autophagic death of HepG2 cells through AMPK activation (Yu et al. 2014). In com-bination studies, the combined use of rapamycin and BBR was found to have a syner-gistic cytotoxic effect through inhibition of the mTOR signaling pathway (Guo et al. 2014). In another study, BBR has been proved as a potential adjuvant agent as this compound in combination with vincristine has been reported for its pro-apoptotic and ER stress-inducing potential in cancer chemotherapy, and it was suggested as a hopeful approach for developing hepatoma therapy by utilizing the combinational

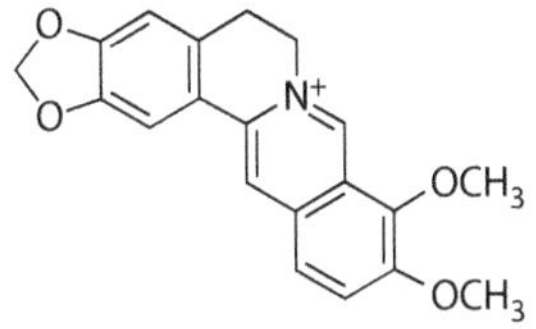

FIGURE 9.8 Chemical structure of BBR ($C_{20}H_{19}NO_5$, CAS. No: 2086-83-1).

effect of vincristine and BBR (Wang et al. 2014). *In vivo,* BBR in combination with S-allyl cysteine (SAC) was studied against DEN and carbon tetrachloride-induced hepatocarcinoma in rats. This BBR + SAC combination treatment for a month significantly inhibited Akt-mediated cell proliferation and inhibited JNK signaling pathway resulting in induction of apoptosis (Sengupta et al. 2014).

In neural glioblastoma multiforme (GBM) cells, BBR is shown to reduce invasiveness, proliferation, and apoptotic induction. Berberine also has profound effects on the metabolic state of GBM cells, leading to high autophagy flux and impaired glycolytic capacity through inhibition of the AMPK/mTOR/ULK1 pathway. Further, *in vivo* study also confirms the anticancer effect of BBR against neural glioblastoma in Athymic mice (Wang et al. 2016b).

In animal studies, BBR has been shown to suppress 20-methylcholanthrene-induced carcinogenesis (Anis et al. 2001) and also 12-O-tetradecanoylphorbol-13-acetate and teleocidin-induced skin tumor promotion (Nishino et al. 1986), and tumor invasion of human lung cancer cells via decreased productions of urokinase-plasminogen activator and MMP-2 (Peng et al. 2006). Tumor invasion requires the expression of MMP-2 and MMP-9, which was also suppressed by BBR and various carcinogens and tumor promoters that have been shown to activate NF-κB. Thus, suppression of NF-κB by BBR as shown here may contribute to its ability to suppress carcinogenesis.

Currently, BBR is in phase I clinical trial for the study against CRC in patients with ulcerative colitis in remission (ClinicalTrials.gov Identifier NCT02365480) and is also in phase II and III clinical trials for its evaluation against colorectal adenomas (ClinicalTrials.gov Identifier NCT02226185).

CAPSAICIN

Capsaicin (trans-8-methyl-*N*-vanillyl-6-nonenamide) is a homovanillic acid derivative responsible for the characteristic pungent feeling of the genus Capsicum (Figure 9.9). Capsaicin (CAP), a major constituent of green and red peppers, has been found as a cancer preventive agent and shows wide applications against various types of cancer (Clark and Lee 2016).

Previous studies have concretely proved the anticancer potential of this compound. In bladder cancer cells, CAP treatment suppresses tumorigenesis by inhibiting its proliferation both *in vitro* and *in vivo.* Capsaicin-induced cell cycle arrest at

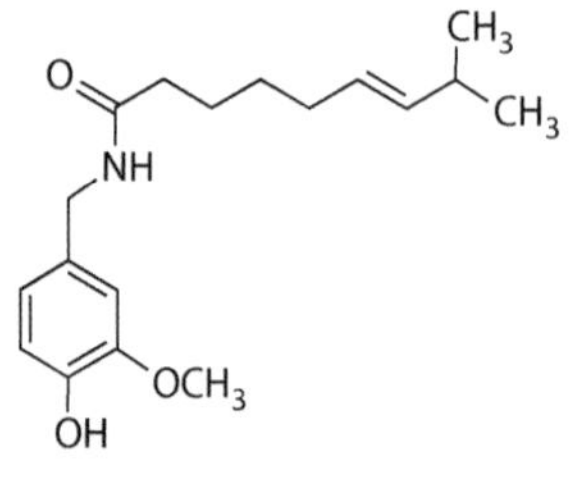

FIGURE 9.9   Chemical structure of CAP ($C_{18}H_{27}NO_3$; CAS. No: 404-86-4).

G0/G1 phase and ROS production and expression of FOXO3a suggest that CAP can inhibit viability and tumorigenesis of bladder cancer possibly via FOXO3a-mediated pathways (Qian et al. 2016). Capsaicin was also shown to inhibit the activation of ERK, thereby reducing the phosphorylation of paxillin and FAK, which leads to decreased cell migration. It inhibits the growth of bladder cancer cells by inhibiting tumor-associated NADH oxidase (tNOX) and SIRT1; thereby reducing proliferation, attenuating migration, and prolonging cell cycle progression (Lin et al. 2016). Capsaicin has been shown exclusively targeting angiogenesis by downregulating VEGF in NSCLC (Chakraborty et al. 2014).

Capsaicin exhibits anticancer activity in HCC cell lines. In this study, CAP induced apoptosis in HepG2 cells. The expression levels of bcl-2 were significantly increased, and with the apoptosis, capsaicin also triggered autophagy in HepG2 cells. Capsaicin has upregulated the STAT3 activity, which contributed to autophagy, and it also triggered ROS generation in hepatoma cells. This study was able to demonstrate that CAP increased the phosphorylation of the signal transducer and activator of transcription 3 (p-STAT3)-dependent autophagy through the generation of ROS signaling pathways in human hepatoma (Chen et al. 2016).

In a renal cancer cell line (786-O), CAP reduced proliferation and induced apoptosis. The apoptosis was found to associate with substantial upregulation of pro-apoptotic genes including c-myc, Bax, and cleaved-caspase-3, -8, and -9, and with downregulation of anti-apoptotic gene Bcl2. CAP treatment also activated P38 and JNK MAPK pathways. Furthermore, CAP significantly slows the growth of 786-O renal cancer xenografts *in vivo* (Liu et al. 2016). More recently, CAP was shown to suppress cell growth by altering histone acetylation in gastric cancer cell lines. This study revealed the important role of hMOF-mediated histone acetylation in CAP-directed anticancer processes, and suggested CAP as a potential drug for use in gastric cancer prevention and therapy (Wang et al. 2016).

In a neural cancer model, CAP was reported to downregulate the expression of the STAT3-regulated genes, such as cyclin D1, Bcl-2, Bcl-xL, survivin, and VEGF, thereby potentiating the apoptotic effects of Velcade and thalidomide in multiple myeloma cells. In an *in vivo* study, intraperitoneal injections of CAP inhibited the growth of human multiple myeloma xenograft tumors in male athymic nu/nu mice (Bhutani et al. 2007).

It has been demonstrated that CAP induces apoptosis in many types of cancer cell lines, including colon adenocarcinoma, pancreatic cancer, hepatocellular carcinoma, prostate cancer, and breast cancer (Bley et al. 2012). Though the molecular mechanism through which CAP induces apoptosis in cancer cells is not completely elucidated, it seems to involve an intracellular calcium increase, generation of ROS, alteration of mitochondrial membrane transition potential, and activation of transcription factors such as NFκB and STATS (Chapa-Oliver and Mejía-Teniente 2016).

In clinical studies, bladder toxicity (cancer) has been screened in those who have been treated with capsaicin over a 5-year period. No pre-malignant or malignant change has been found in biopsies of patients who had repeated capsaicin instillations for up to 5 years. This study shows the safety of CAP for its chronic treatment in cancer patients (Dasgupta et al. 1998). A topical capsaicin cream is found to relieve postsurgical neuropathic pain in cancer patients, despite some toxicity such as skin burning and skin redness (Ellison et al. 1997). Since then several clinical studies were performed with CAP to explore its anticancer efficacy in cancer patients. In a phase III clinical trial, CAP lozenges were studied in head and neck cancer patients with mucositis caused by radiation therapy (ClinicalTrials. gov Identifier NCT00003610). Currently, CAP is in phase II clinical trial for its chemopreventive effect against prostate cancer (ClinicalTrials.gov Identifier NCT00130962).

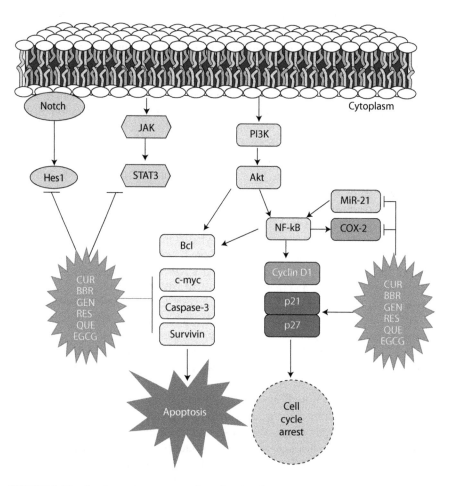

FIGURE 9.10   A schematic representation of molecular targets against various cancers regulated by plant-derived compounds. CUR: curcumin; BBR: berberine; GEN: genistein; RES: resveratrol; QUE: quercetin; EGCG: epigallocatechingallate; NF-κB: Nuclear factor kappa-light-chain-enhancer of activated B cells; COX2: Cyclooxygenase 2; Hes-1: Transcription factor HES1; Akt: Protein kinase B; Stat3: Signal transducer and activator of transcription 3; PI3K: phosphatidylinositol 3-kinase; Notch-1: Neurogenic locus notch homolog protein-1; c-myc: C-mycproto-oncogene; Jak: Janus kinase. p21: Cyclin-dependent kinase inhibitor; p27: Cyclin-dependent kinase inhibitor; BCL: B-cell lymphoma.

CONCLUSION

The compounds reviewed in this chapter are natural products with promising anti-cancer activity. Here we discussed some of the important plant-derived drugs that have been recently studied for their potential correlation with anticancer effects. Unfortunately, it was not possible to cite all of the important works that have been reported in this broad and active field. Undoubtedly, all of these compounds act against cancer cell proliferation, metastasis, and angiogenesis, and cause cellular death through induction of apoptosis, cell cycle arrest mediated through various molecular mechanisms (Figure 9.10). Overwhelming evidence from myriad *in vitro* and *in vivo* studies has concretely demonstrated the potential of these plant-derived alkaloids, flavonoids, and polyphenolic compounds to treat and prevent various cancer types. A summary of preclinical data presented in this chapter shows the remarkable efficacy of these agents against various cancer cells *in vitro* with molecular targets ranging from cell cycle arrest, induction of apoptosis, autophagy, induction of ROS, etc. In addition, animal experiments including xenograft models support that the phytomedicines have clinical activity in human cancer therapy. Nevertheless, many protocols and endpoints chosen for the phase I and II studies of several cancer therapies are of some concern and make interpretation of data quite difficult. Doses and concentrations of the drug/substance used in these studies have mostly been empirically derived and rarely tested systematically, and the manufacturing and preparation have not been standardized. In addition, the association of these plant-derived compounds with current standard anticancer drugs has demonstrated synergistic effect useful to improve cancer treatment. However, the use of herbal drugs that were not previously evaluated for their safety may expose patients to a wide spectrum of risks that may range from undertreatment due to the delay in using official medicine treatment, to toxicities derived both as a direct consequence of the alternative molecule or because of drug interaction with conventional treatments. Further, *in vitro* and *in vivo* studies are warranted to shed more light on the potential clinical utility of these compounds against various cancers. Several clinical trials aim to investigate the beneficial effects of these plant-derived compounds in humans and have produced promising results. Hence, the use of these plant-derived phytochemicals seems to contribute to anticancer therapy due to their ability to induce apoptosis and cell cycle arrest and other molecular mechanisms discussed elsewhere in this chapter.

REFERENCES

Abdel-Lateef E, Mahmoud F, Hammam O, El-Ahwany E, El-Wakil E, Kandil S, et al. Bioactive chemical constituents of *Curcuma longa* L. rhizomes extract inhibit the growth of human hepatoma cell line (HepG2). *Acta Pharm.* 2016;66(3):387–398.

Abouzied MM, Eltahir HM, Abdel Aziz MA, Ahmed NS, Abd El-Ghany AA, Abd El-Aziz EA, Abd El-Aziz HO. Curcumin ameliorate DENA-induced HCC via modulating TGF-β, AKT, and caspase-3 expression in experimental rat model. *Tumour Biol.* 2015;36(3):1763–1771.

Aggarwal BB, Bhardwaj A, Aggarwal RS, Seeram NP, Shishodia S, Takada Y. Role of resveratrol in prevention and therapy of cancer: Preclinical and clinical studies. *Anticancer Res.* 2004;24(5A):2783–2840.

Akhtar R, Ali M, Mahmood S, Sanyal SN. Anti-proliferative action of silibinin on human colon adenomatous cancer HT-29 cells. *Nutr Hosp.* 2014;29(2):388–392.

Alayev A, Berger SM, Holz MK. Resveratrol as a novel treatment for diseases with mTOR pathway hyperactivation. *Ann N Y Acad Sci.* 2015;1348(1):116–123.

Anis KV, Rajeshkumar NV, Kuttan R. Inhibition of chemical carcinogenesis by berberine in rats and mice. *J Pharm Pharmacol.* 2001;53:763–768.

Banerjee S, Li Y, Wang Z, Sarkar FH. Multi-targeted therapy of cancer by genistein. *Cancer Lett.* 2008;269:226–242.

Bayram D, Çetin ES, Kara M, Özgöçmen M, Candan IA. The apoptotic effects of silibinin on MDA-MB-231 and MCF-7 human breast carcinoma cells. *Hum Exp Toxicol.* 2017;36(6):573–586.

Bhatia N, Zhao J, Wolf DM, Agarwal R. Inhibition of human carcinoma cell growth and DNA synthesis by silibinin, an active constituent of milk thistle: Comparison with silymarin. *Cancer Lett.* 1999;147(1–2):77–84.

Bhutani M, Pathak AK, Nair AS, Kunnumakkara AB, Guha S, Sethi G, Aggarwal BB. Capsaicin is a novel blocker of constitutive and interleukin-6-inducible STAT3 activation. *Clin Cancer Res.* 2007;13(10):3024–3032.

Blanquer-Rosselló MD, Hernández-López R, Roca P, Oliver J, Valle A. Resveratrol induces mitochondrial respiration and apoptosis in SW620 colon cancer cells. *Biochim Biophys Acta.* 2017;1861(2):431–440.

Bley K, Boorman G, Mohammad B, McKenzie D, Babbar S. A comprehensive review of the carcinogenic and anticarcinogenic potential of capsaicin. *Toxicol Pathol.* 2012;40(6):847–873.

Bosch-Barrera J, Menendez JA. Silibinin and STAT3: A natural way of targeting transcription factors for cancer therapy. *Cancer Treat Rev.* 2015;41(6):540–546.

Bozza C, Agostinetto E, Gerratana L, Puglisi F. Complementary and alternative medicine in oncology. *Recenti Prog Med.* 2015;106(12):601–607.

Burstein HJ, Gelber S, Guadagnoli E, Weeks JC. Use of alternative medicine by women with early-stage breast cancer. *N Engl J Med.* 1999;340(22):1733–1739.

Carafa V, Rotili D, Forgione M, Cuomo F, Serretiello E, Hailu GS, et al. Sirtuin functions and modulation: From chemistry to the clinic. *Clin Epigenetics.* 2016;8:61.

Carelle N, Piotto E, Bellanger A, Germanaud J, Thuillier A, Khayat D. Changing patient perceptions of the side effects of cancer chemotherapy. *Cancer.* 2002;95(1):155–163.

Carroll RE, Benya RV, Turgeon DK, Vareed S, Neuman M, Rodriguez L, et al. Phase IIa clinical trial of curcumin for the prevention of colorectal neoplasia. *Cancer Prev Res (Phila).* 2011;4(3):354–364.

Chakraborty S, Adhikary A, Mazumdar M, Mukherjee S, Bhattacharjee P, Guha D, et al. Capsaicin-induced activation of p53-SMAR1 auto-regulatory loop down-regulates VEGF in non-small cell lung cancer to restrain angiogenesis. *PLoS One.* 2014;9(6):e99743.

Chan CY, Lien CH, Lee MF, Huang CY. Quercetin suppresses cellular migration and invasion in human head and neck squamous cell carcinoma (HNSCC). *Biomedicine (Taipei).* 2016;6(3):10–15.

Chapa-Oliver AM, Mejía-Teniente L. Capsaicin: From plants to a cancer-suppressing agent. *Molecules.* 2016;21(8):931.

Chen XW, Di YM, Zhang J, Zhou ZW, Li CG, Zhou SF. Interaction of herbal compounds with biological targets: A case study with berberine. *Sci World J.* 2012;2012:708292.

Chen X, Tan M, Xie Z, Feng B, Zhao Z, Yang K, et al. Inhibiting ROS-STAT3-dependent autophagy enhanced capsaicin-induced apoptosis in human hepatocellular carcinoma cells. *Free Radic Res.* 2016;50(7):744–755.

Chin YT, Hsieh MT, Yang SH, Tsai PW, Wang SH, Wang CC, et al. Anti-proliferative and gene expression actions of resveratrol in breast cancer cells *in vitro. Oncotarget.* 2014;5(24):12891–12907.

Chung MY, Lim TG, Lee KW. Molecular mechanisms of chemopreventive phytochemicals against gastroenterological cancer development. *World J Gastroenterol.* 2013;19:984–993.

Clark R, Lee SH. Anticancer properties of capsaicin against human cancer. *Anticancer Res.* 2016;36(3):837–843.

Dai XZ, Yin HT, Sun LF, Hu X, Zhou C, Zhou Y, et al. Potential therapeutic efficacy of curcumin in liver cancer. *Asian Pac J Cancer Prev.* 2013;14(6):3855–3859.

Dasgupta P, Chandiramani V, Parkinson MC, Beckett A, Fowler CJ. Treating the human bladder with capsaicin: Is it safe? *Eur Urol.* 1998;33(1):28–31.

Dastjerdi MN, Kavoosi F, Valiani A, Esfandiari E, Sanaei M, Sobhanian S, et al. Inhibitory effect of genistein on PLC/PRF5 hepatocellular carcinoma cell line. *Int J Prev Med.* 2015;6:54.

Deep G, Oberlies NH, Kroll DJ, Agarwal R. Identifying the differential effects of silymarin constituents on cell growth and cell cycle regulatory molecules in human prostate cancer cells. *Int J Cancer.* 2008;123(1):41–50.

Deguchi H, Fujii T, Nakagawa S, Koga T, Shirouzu K. Analysis of cell growth inhibitory effects of catechin through MAPK in human breast cancer cell line T47D. *Int J Oncol.* 2002;21(6):1301–1305.

Dhandayuthapani S, Marimuthu P, Hörmann V, Kumi-Diaka J, Rathinavelu A. Induction of apoptosis in HeLa cells via caspase activation by resveratrol and genistein. *J Med Food.* 2013;16(2):139–146.

Dhillon N, Aggarwal BB, Newman RA, Wolff RA, Kunnumakkara AB, Abbruzzese JL, et al. Phase II trial of curcumin in patients with advanced pancreatic cancer. *Clin Cancer Res.* 2008;14(14):4491–4499.

Du GJ, Zhang Z, Wen XD, Yu C, Calway T, Yuan CS, Wang CZ. Epigallocatechin Gallate (EGCG) is the most effective cancer chemopreventive polyphenol in green tea. *Nutrients.* 2012;4(11):1679–1691.

Ellison N, Loprinzi CL, Kugler J, Hatfield AK, Miser A, Sloan JA, et al. Phase III placebo-controlled trial of capsaicin cream in the management of surgical neuropathic pain in cancer patients. *J Clin Oncol.* 1997;15(8):2974–2980.

Epelbaum R, Schaffer M, Vizel B, Badmaev V, Bar-Sela G. Curcumin and gemcitabine in patients with advanced pancreatic cancer. *Nutr Cancer.* 2010;62(8):1137–1141.

Ezhilarasan D, Karthikeyan S. Silibinin alleviates N-nitrosodimethylamine-induced glutathione dysregulation and hepatotoxicity in rats. *Chin J Nat Med.* 2016;14(1):40–47.

Ezhilarasan D, Karthikeyan S, Vivekanandan P. Ameliorative effect of silibinin against N-nitrosodimethylamine-induced hepatic fibrosis in rats. *Environ Toxicol Pharmacol.* 2012;34(3):1004–1013.

Ezhilarasan D, Evraerts J, Brice S, Buc-Calderon P, Karthikeyan S, Sokal E, Najimi M. Silibinin inhibits proliferation and migration of human hepatic stellate LX-2 cells. *J Clin Exp Hepatol.* 2016;6(3):167–174.

Fang MZ, Wang Y, Ai N, Hou Z, Sun Y, Lu H, et al. Tea polyphenol (-)-epigallocatechin-3-gallateinhibits DNA methyltransferase and reactivates methylation-silenced genes in cancer cell lines. *Cancer Res.* 2003;63:7563–7570.

Fang JY, Tsai TH, Lin YY, Wong WW, Wang MN, Huang JF. Transdermal delivery of tea catechins and theophylline enhanced by terpenes: A mechanistic study. *Biol Pharm Bull.* 2007;30:343–349.

Feng Y, Zhou J, Jiang Y. Resveratrol in lung cancer—A systematic review. *J BUON.* 2016;21(4):950–953.

Ferlay J, Soerjomataram I, Dikshit R, Eser S, Mathers C, Rebelo M, et al. Cancer incidence and mortality worldwide: Sources, methods and major patterns in GLOBOCAN 2012. *Int J Cancer.* 2015;136(5):E359–E386.

Flaig TW, Gustafson DL, Su LJ, Zirrolli JA, Crighton F, Harrison GS, et al. A phase I and pharmacokinetic study of silybin-phytosome in prostate cancer patients. *Invest New Drugs*. 2007;25(2):139–146.

Flaig TW, Glodé M, Gustafson D, van Bokhoven A, Tao Y, Wilson S, et al. A study of high-dose oral silybin-phytosome followed by prostatectomy in patients with localized prostate cancer. *Prostate*. 2010;70(8):848–855.

Gallardo M, Calaf GM. Curcumin inhibits invasive capabilities through epithelial mesenchymal transition in breast cancer cell lines. *Int J Oncol*. 2016;49(3):1019–1027.

García-Maceira P, Mateo J. Silibinin inhibits hypoxia-inducible factor-1alpha and TOR/p70S6 K/4E-BP1 signalling pathway in human cervical and hepatoma cancer cells: Implications for anticancer therapy. *Oncogene*. 2009;28(3):313–324.

GLOBOCAN. 2012. Available from: http://globocan.iarc.fr/Default.aspx. Accessed May 12, 2016.

Gong L, Li Y, Nedeljkovic-Kurepa A, Sarkar FH. Inactivation of NF-κB by genistein is mediated via Akt signaling pathway in breast cancer cells. *Oncogene*. 2003;22:4702–4709.

Guo N, Yan A, Gao X, Chen Y, He X, Hu Z, et al. Berberine sensitizes rapamycin-mediated human hepatoma cell death *in vitro*. *Mol Med Rep*. 2014;10(6):3132–3138.

Han G, Xia J, Gao J, Inagaki Y, Tang W, Kokudo N. Antitumor effects and cellular mechanisms of resveratrol. *Drug Discov Ther*. 2015;9(1):1–12.

Han M, Song Y, Zhang X. Quercetin suppresses the migration and invasion in human colon cancer caco-2 cells through regulating toll-like receptor 4/nuclear factor-kappa B pathway. *Pharmacogn Mag*. 2016;12(Suppl 2):S237–S244.

He ZY, Shi CB, Wen H, Li FL, Wang BL, Wang J. Upregulation of p53 expression in patients with colorectal cancer by administration of curcumin. *Cancer Invest*. 2011;29(3):208–213.

Hoh C, Boocock D, Marczylo T, Singh R, Berry DP, Dennison AR, et al. Pilot study of oral silibinin, a putative chemopreventive agent, in colorectal cancer patients: Silibinin levels in plasma, colorectum, and liver and their pharmacodynamic consequences. *Clin Cancer Res*. 2006;12(9):2944–2950.

Hörmann V, Kumi-Diaka J, Durity M, Rathinavelu A. Anticancer activities of genistein-topotecan combination in prostate cancer cells. *J Cell Mol Med*. 2012;16(11):2631–2636.

Hsieh CY, Santell RC, Haslam SZ, Helferich WG. Estrogenic effects of genistein on the growth of estrogen receptor-positive human breast cancer (MCF-7) cells *in vitro* and *in vivo*. *Cancer Res*. 1998;58:3833–3838.

Hsu WH, Hsieh YS, Kuo HC, Teng CY, Huang HI, Wang CJ, et al. Berberine induces apoptosis in SW620 human colonic carcinoma cells through generation of reactive oxygen species and activation of JNK/p38 MAPK and FasL. *Arch Toxicol*. 2007;81(10):719–728.

Hu Y, Wang S, Wu X, Zhang J, Chen R, Chen M, Wang Y. Chinese herbal medicine-derived compounds for cancer therapy: A focus on hepatocellular carcinoma. *J Ethnopharmacol*. 2013;149(3):601–612.

Hu YQ, Wang J, Wu JH. Administration of resveratrol enhances cell-cycle arrest followed by apoptosis in DMBA-induced skin carcinogenesis in male Wistar rats. *Eur Rev Med Pharmacol Sci*. 2016;20(13):2935–2946.

Hung LM, Chen JK, Huang SS, Lee RS, Su MJ. Cardioprotective effect of resveratrol, a natural antioxidant derived from grapes. *Cardiovasc Res*. 2000;47(3):549–555.

Hwang JT, Ha J, Park IJ, Lee SK, Baik HW, Kim YM, Park OJ. Apoptotic effect of EGCG in HT-29 colon cancer cells via AMPK signal pathway. *Cancer Lett*. 2007;247(1):115–121.

Iizasa H, Nanbo A, Nishikawa J, Jinushi M, Yoshiyama H. Epstein–Barr virus (EBV)-associated gastric carcinoma. *Viruses*. 2012;4(12):3420–3439.

Inoue M, Robien K, Wang R, Van Den Berg DJ, Koh WP, Yu MC. Green tea intake, MTHFR/TYMS genotype and breast cancer risk: The Singapore Chinese Health Study. *Carcinogenesis*. 2008;29(10):1967–1972.

Irimie AI, Braicu C, Zanoaga O, Pileczki V, Gherman C, Berindan-Neagoe I, Campian RS. Epigallocatechin-3-gallate suppresses cell proliferation and promotes apoptosis and autophagy in oral cancer SSC-4 cells. *Onco Targets Ther.* 2015;8:461–470.

Jesudas B, Raghu R, Bhavani G, Ezhilarasan D, Karthikeyan S. Ethanol enhances lamivudine-induced liver toxicity: Investigation on hepatoprotective properties of silibinin-phosphotidyl choline complex in rats. *Biomedicine.* 2016;36(1):128–137.

Jang M, Cai L, Udeani GO, Slowing KV, Thomas CF, Beecher CW, et al. Cancer chemopreventive activity of resveratrol, a natural product derived from grapes. *Science.* 1997;275(5297):218–220.

Ju YH, Allred KF, Allred CD, Helferich WG. Genistein stimulates growth of human breast cancer cells in a novel, postmenopausal animal model, with low plasma estradiol concentrations. *Carcinogenesis.* 2006;27:1292–1299.

Kadasa NM, Abdallah H, Afifi M, Gowayed S. Hepatoprotective effects of curcumin against diethyl nitrosamine induced hepatotoxicity in albino rats. *Asian Pac J Cancer Prev.* 2015;16(1):103–108.

Kaur S, Greaves P, Cooke DN, Edwards R, Steward WP, Gescher AJ, Marczylo TH. Breast cancer prevention by green tea catechins and black tea theaflavins in the C3(1) SV40T, t antigen transgenic mouse model is accompanied by increased apoptosis and a decrease in oxidative DNA adducts. *J Agric Food Chem.* 2007;55(9):3378–3385.

Kavoosi F, Dastjerdi MN, Valiani A, Esfandiari E, Sanaei M, Hakemi MG. Genistein potentiates the effect of 17-beta estradiol on human hepatocellular carcinoma cell line. *Adv Biomed Res.* 2016;5:133.

Khan N, Afaq F, Saleem M, Ahmad N, Mukhtar H. Targeting multiple signaling pathways by green tea polyphenol (-)-epigallocatechin-3-gallate. *Cancer Res.* 2006;66:2500–2505.

Kim SH, Kim KY, Yu SN, Seo YK, Chun SS, Yu HS, Ahn SC. Silibinin induces mitochondrial NOX4-mediated endoplasmic reticulum stress response and its subsequent apoptosis. *BMC Cancer.* 2016;16:452.

Klinger NV, Mittal S. Therapeutic potential of curcumin for the treatment of brain tumors. *Oxid Med Cell Longev.* 2016;2016:9324085.

Kumar N, Titus-Ernstoff L, Newcomb PA, Trentham-Dietz A, Anic G, Egan KM. Tea consumption and risk of breast cancer. *Cancer Epidemiol Biomarkers Prev.* 2009;18(1):341–345.

Kumar D, Basu S, Parija L, Rout D, Manna S, Dandapat J, Debata PR. Curcumin and ellagic acid synergistically induce ROS generation, DNA damage, p53 accumulation and apoptosis in HeLa cervical carcinoma cells. *Biomed Pharmacother.* 2016;81:31–37.

Kuo CL, Chi CW, Liu TY. Modulation of apoptosis by berberine through inhibition of cyclooxygenase-2 and Mcl-1 expression in oral cancer cells. *In Vivo.* 2005;19(1):247–252.

Kuo HP, Chuang TC, Yeh MH, Hsu SC, Way TD, Chen PY, et al. Growth suppression of HER2-overexpressing breast cancer cells by berberine via modulation of the HER2/PI3K/Akt signaling pathway. *J Agric Food Chem.* 2011;59(15):8216–8224.

Lamartiniere CA, Zhang JX, Cotroneo MS. Genistein studies in rats: Potential for breast cancer prevention and reproductive and developmental toxicity. *Am J Clin Nutr.* 1998;68:1400S–1405S.

Lazzeroni M, Guerrieri-Gonzaga A, Gandini S, Johansson H, Serrano D, Cazzaniga M, et al. A presurgical study of oral silybin-phosphatidylcholine in patients with early breast cancer. *Cancer Prev Res (Phila).* 2016;9(1):89–95.

Lee HH, Lee S, Shin YS, Cho M, Kang H, Cho H. Anti-cancer effect of quercetin in xenograft models with EBV-associated human gastric carcinoma. *Molecules.* 2016b;21(10):1286.

Lee JW, Park S, Kim SY, Um SH, Moon EY. Curcumin hampers the antitumor effect of vinblastine via the inhibition of microtubule dynamics and mitochondrial membrane potential in HeLa cervical cancer cells. *Phytomedicine.* 2016a;23(7):705–713.

Li Y, Ahmed F, Ali S, Philip PA, Kucuk O, Sarkar FH. Inactivation of nuclear factor kappa B by soy isoflavone genistein contributes to increased apoptosis induced by chemotherapeutic agents in human cancer cells. *Cancer Res.* 2005;65:6934–6942.

Li Y, Zhu W, Li J, Liu M, Wei M. Resveratrol suppresses the STAT3 signaling pathway and inhibits proliferation of high glucose-exposed HepG2 cells partly through SIRT1. *Oncol Rep.* 2013;30(6):2820–2828.

Li J, Li O, Kan M, Zhang M, Shao D, Pan Y, et al. Berberine induces apoptosis by suppressing the arachidonic acid metabolic pathway in hepatocellular carcinoma. *Mol Med Rep.* 2015;12(3):4572–4577.

Li M, Tse LA, Chan WC, Kwok CH, Leung SL, Wu C, et al. Evaluation of breast cancer risk associated with tea consumption by menopausal and estrogen receptor status among Chinese women in Hong Kong. *Cancer Epidemiol.* 2016;40:73–78.

Liang Q, Wang C, Li B, Zhang AH. Metabolic fingerprinting to understand therapeutic effects and mechanisms of silybin on acute liver damage in rat. *Pharmacogn Mag.* 2015;11(43):586–593.

Lim W, Yang C, Park S, Bazer FW, Song G. Inhibitory effects of quercetin on progression of human choriocarcinoma cells are mediated through PI3K/AKT and MAPK signal transduction cascades. *J Cell Physiol.* 2017;232(6):1428–1440.

Lin JN, Lin VC, Rau KM, Shieh PC, Kuo DH, Shieh JC, et al. Resveratrol modulates tumor cell proliferation and protein translation via SIRT1-dependent AMPK activation. *J Agric Food Chem.* 2010;58(3):1584–1592.

Lin MH, Lee YH, Cheng HL, Chen HY, Jhuang FH, Chueh PJ. Capsaicin inhibits multiple bladder cancer cell phenotypes by inhibiting tumor-associated NADH oxidase (tNOX) and Sirtuin1 (SIRT1). *Molecules.* 2016;21(7):849.

Liu Y, Zhang YM, Song DF, Cui HB. Effect of apoptosis in human breast cancer cells and its probable mechanisms by genistein. *Wei Sheng Yan Jiu.* 2005;34:67–69.

Liu T, Wang G, Tao H, Yang Z, Wang Y, Meng Z, et al. Capsaicin mediates caspases activation and induces apoptosis through P38 and JNK MAPK pathways in human renal carcinoma. *BMC Cancer.* 2016;16(1):790.

Lopes-Rodrigues V, Sousa E, Vasconcelos MH. Curcumin as a modulator of P-glycoprotein in cancer: Challenges and perspectives. *Pharmaceuticals (Basel).* 2016;9(4):71.

Luo Y, Wang SX, Zhou ZQ, Wang Z, Zhang YG, Zhang Y, Zhao P. Apoptotic effect of genistein on human colon cancer cells via inhibiting the nuclear factor-kappa B (NF-κB) pathway. *Tumour Biol.* 2014;35(11):11483–11488.

Luther DJ, Ohanyan V, Shamhart PE, Hodnichak CM, Sisakian H, Booth TD, et al. Chemopreventive doses of resveratrol do not produce cardiotoxicity in a rodent model of hepatocellular carcinoma. *Invest New Drugs.* 2011;29(2):380–391.

Mahammedi H, Planchat E, Pouget M, Durando X, Curé H, Guy L, et al. The new combination docetaxel, prednisone and curcumin in patients with castration-resistant prostate cancer: A pilot phase ii study. *Oncology.* 2016;90(2):69–78.

Mai Z, Blackburn GL, Zhou JR. Genistein sensitizes inhibitory effect of tamoxifen on the growth of estrogen receptor-positive and HER2-overexpressing human breast cancer cells. *Mol Carcinog.* 2007;46:534–542.

Mateen S, Tyagi A, Agarwal C, Singh RP, Agarwal R. Silibinin inhibits human non-small cell lung cancer cell growth through cell-cycle arrest by modulating expression and function of key cell-cycle regulators. *Mol Carcinog.* 2010;49(3):247–258.

Mohammadi A, Mansoori B, Baradaran B. The role of microRNAs in colorectal cancer. *Biomed Pharmacother.* 2016;84:705–713.

Montgomery A, Adeyeni T, San K, Heuertz RM, Ezekiel UR. Curcumin sensitizes silymarin to exert synergistic anticancer activity in colon cancer cells. *J Cancer.* 2016;7(10):1250–1257.

Mukherjee S, Dudley JI, Das DK. Dose-dependency of resveratrol in providing health benefits. *Dose Response*. 2010;8(4):478–500.

Muthoosamy K, Abubakar IB, Bai RG, Loh HS, Manickam S. Exceedingly higher co-loading of curcumin and paclitaxel onto polymer-functionalized reduced graphene oxide for highly potent synergistic anticancer treatment. *Sci Rep*. 2016;6:32808.

Nandakumar V, Vaid M, Katiyar SK. (-)-Epigallocatechin-3-gallate reactivates silenced tumor suppressor genes, Cip1/p21 and p16INK4a, by reducing DNA methylation and increasing histones acetylation in human skin cancer cells. *Carcinogenesis*. 2011;32(4):537–544.

National Cancer Institute. Available from: https://www.cancer.gov/types/common-cancers. Accessed November 28, 2016.

Nishino H, Kitagawa K, Fujiki H, Iwashima A. Berberine sulfate inhibits tumor-promoting activity of teleocidin in two-stage carcinogenesis on mouse skin. *Oncology*. 1986;43(2):131–134.

Nwaeburu CC, Bauer N, Zhao Z, Abukiwan A, Gladkich J, Benner A, Herr I. Up-regulation of microRNA Let-7c by quercetin inhibits pancreatic cancer progression by activation of Numbl. *Oncotarget*. 2016;7(36):58367–58380.

Oliveira MR, Nabavi SF, Daglia M, Rastrelli L, Nabavi SM. Epigallocatechin gallate and mitochondria—A story of life and death. *Pharmacol Res*. 2016;104:70–75.

Pavan AR, Silva GD, Jornada DH, Chiba DE, Fernandes GF, Man Chin C, Dos Santos JL. Unraveling the anticancer effect of curcumin and resveratrol. *Nutrients*. 2016;8(11):628.

Peng PL, Hsieh YS, Wang CJ, Hsu JL, Chou FP. Inhibitory effect of berberine on the invasion of human lung cancer cells via decreased productions of urokinase-plasminogen activator and matrix metalloproteinase-2. *Toxicol Appl Pharmacol*. 2006;214(1):8–15.

Pirouzpanah MB, Sabzichi M, Pirouzpanah S, Chavoshi H, Samadi N. Silibilin-induces apoptosis in breast cancer cells by modulating p53, p21, Bak and Bcl-XL pathways. *Asian Pac J Cancer Prev*. 2015;16(5):2087–2092.

Prasad S, Gupta SC, Tyagi AK, Aggarwal BB. Curcumin, a component of golden spice: From bedside to bench and back. *Biotechnol Adv*. 2014;32:1053–1064.

Prasain JK, Rajbhandari R, Keeton AB, Piazza GA, Barnes S. Metabolism and growth inhibitory activity of cranberry derived flavonoids in bladder cancer cells. *Food Funct*. 2016;7(9):4012–4019.

Qian K, Wang G, Cao R, Liu T, Qian G, Guan X, et al. Capsaicin suppresses cell proliferation, induces cell cycle arrest and ROS production in bladder cancer cells through FOXO3a-mediated pathways. *Molecules*. 2016;21(10):1406.

Quispe-Soto ET, Calaf GM. Effect of curcumin and paclitaxel on breast carcinogenesis. *Int J Oncol*. 2016;49(6):2569–2577.

Raghu R, Jesudas B, Bhavani G, Ezhilarasan D, Karthikeyan S. Silibinin mitigates zidovudine-induced hepatocellular degenerative changes, oxidative stress and hyperlipidaemia in rats. *Hum Exp Toxicol*. 2015;34(11):1031–1042.

Rai G, Mishra S, Suman S, Shukla Y. Resveratrol improves the anticancer effects of doxorubicin *in vitro* and *in vivo* models: A mechanistic insight. *Phytomedicine*. 2016;23(3):233–242.

Ranjan AP, Mukerjee A, Gdowski A, Helson L, Bouchard A, Majeed M, Vishwanatha JK. Curcumin-er prolonged subcutaneous delivery for the treatment of non-small cell lung cancer. *J Biomed Nanotechnol*. 2016;12(4):679–688.

Rathore K, Choudhary S, Odoi A, Wang HC. Green tea catechin intervention of reactive oxygen species-mediated ERK pathway activation and chronically induced breast cell carcinogenesis. *Carcinogenesis*. 2012;33(1):174–183.

Reddivari L, Charepalli V, Radhakrishnan S, Vadde R, Elias RJ, Lambert JD, Vanamala JK. Grape compounds suppress colon cancer stem cells in vitro and in a rodent model of colon carcinogenesis. *BMC Complement Altern Med*. 2016;16:278.

Rivera Rivera A, Castillo-Pichardo L, Gerena Y, Dharmawardhane S. Anti-breast cancer potential of quercetin via the Akt/AMPK/mammalian target of rapamycin (mTOR) signaling cascade. *PLoS One.* 2016;11(6):e0157251.

Roh T, Kim SW, Moon SH, Nam MJ. Genistein induces apoptosis by down-regulating thioredoxin-1 in human hepatocellular carcinoma SNU-449 cells. *Food Chem Toxicol.* 2016;97:127–134.

Ruch RJ, Cheng SJ, Klaunig JE. Prevention of cytotoxicity and inhibition of intercellular communication by antioxidant catechins isolated from Chinese green tea. *Carcinogenesis.* 1989;10(6):1003–1008.

Russo M, Russo GL, Daglia M, Kasi PD, Ravi S, Nabavi SF, Nabavi SM. Understanding genistein in cancer: The "good" and the "bad" effects: A review. *Food Chem.* 2016;196:589–600.

Ryu MJ, Cho M, Song JY, Yun YS, Choi IW, Kim DE, et al. Natural derivatives of curcumin attenuate the Wnt/ß-catenin pathway through down-regulation of the transcriptional coactivator p300. *Biochem Biophys Res Commun.* 2008;377:1304–1308.

Sarkar A, Ghosh S, Chowdhury S, Pandey B, Sil PC. Targeted delivery of quercetin loaded mesoporous silica nanoparticles to the breast cancer cells. *Biochim Biophys Acta.* 2016;1860(10):2065–2075.

Satoh H, Nishikawa K, Suzuki K, Asano R, Virgona N, Ichikawa T, et al. Genistein, a soy isoflavone, enhances necrotic like cell death in a breast cancer cell treated with a chemotherapeutic agent. *Res Commun Mol Pathol Pharmacol.* 2003;113–114:149–158.

Sekar V, Anandasadagopan SK, Ganapasam S. Genistein regulates tumor microenvironment and exhibits anticancer effect in dimethyl hydrazine-induced experimental colon carcinogenesis. *Biofactors.* 2016;42(6):623–637.

Sengupta D, Chowdhury KD, Sarkar A, Paul S, Sadhukhan GC. Berberine and S allyl cysteine mediated amelioration of DEN + CCl4 induced hepatocarcinoma. *Biochim Biophys Acta.* 2014;1840(1):219–244.

Shang HS, Chang CH, Chou YR, Yeh MY, Au MK, Lu HF, et al. Curcumin causes DNA damage and affects associated protein expression in HeLa human cervical cancer cells. *Oncol Rep.* 2016;36(4):2207–2215.

Shin SB, Woo SU, Chin YW, Jang YJ, Yim H. Sensitivity of TP53-mutated cancer cells to the phytoestrogen genistein is associated with direct inhibition of plk1 activity. *J Cell Physiol.* 2017;232(10):2818–2828.

Siegel AB, Narayan R, Rodriguez R, Goyal A, Jacobson JS, Kelly K, et al. A phase I dose-finding study of silybin phosphatidylcholine (milk thistle) in patients with advanced hepatocellular carcinoma. *Integr Cancer Ther.* 2014;13(1):46–43.

Siegel RL, Miller KD, Jemal A. Cancer statistics, 2016. *CA Cancer J Clin.* 2016;66(1):7–30.

Singh BN, Shankar S, Srivastava RK. Green tea catechin, epigallocatechin-3-gallate (EGCG): Mechanisms, perspectives and clinical applications. *Biochem Pharmacol.* 2011;82(12):1807–1821.

Sugio A, Iwasaki M, Habata S, Mariya T, Suzuki M, Osogami H, et al. BAG3 upregulates Mcl-1 through downregulation of miR-29b to induce anticancer drug resistance in ovarian cancer. *Gynecol Oncol.* 2014;134(3):615–623.

Thacker PC, Karunagaran D. Curcumin and emodin down-regulate TGF-β signaling pathway in human cervical cancer cells. *PLoS One.* 2015;10(3):e0120045.

Thorburn A, Thamm DH, Gustafson DL. Autophagy and cancer therapy. *Mol Pharmacol.* 2014;85(6):830–838.

Thun MJ, DeLancey JO, Center MM, Jemal A, Ward EM. The global burden of cancer: Priorities for prevention. *Carcinogenesis.* 2010;31(1):100–110.

Tsai ML, Lai CS, Chang YH, Chen WJ, Ho CT, Pan MH. Pterostilbene, a natural analogue of resveratrol, potently inhibits 7,12-dimethylbenz[a]anthracene (DMBA)/12-O-tetradecanoylphorbol-13-acetate (TPA)-induced mouse skin carcinogenesis. *Food Funct.* 2012;3(11):1185–1194.

Tsai CC, Chuang TW, Chen LJ, Niu HS, Chung KM, Cheng JT, Lin KC. Increase in apoptosis by combination of metformin with silibinin in human colorectal cancer cells. *World J Gastroenterol.* 2015;21(14):4169–4177.

Tyagi AK, Singh RP, Agarwal C, Chan DC, Agarwal R. Silibinin strongly synergizes human prostate carcinoma DU145 cells to doxorubicin-induced growth Inhibition, G2-M arrest, and apoptosis. *Clin Cancer Res.* 2002;8(11):3512–3519.

Unlu A, Nayir E, Dogukan Kalenderoglu M, Kirca O, Ozdogan M. Curcumin (Turmeric) and cancer. *J BUON.* 2016;21(5):1050–1060.

Van Aller GS, Carson JD, Tang W, Peng H, Zhao L, Copeland RA, et al. Epigallocatechin gallate (EGCG), a major component of green tea, is a dual phosphoinositide-3-kinase/mTOR inhibitor. *Biochem Biophys Res Commun.* 2011;406(2):194–199.

Vanamala J, Reddivari L, Radhakrishnan S, Tarver C. Resveratrol suppresses IGF-1 induced human colon cancer cell proliferation and elevates apoptosis via suppression of IGF-1R/Wnt and activation of p53 signaling pathways. *BMC Cancer.* 2010;10:238.

Varinska L, Gal P, Mojzisova G, Mirossay L, Mojzis J. Soy and breast cancer: Focus on angiogenesis. *Int J Mol Sci.* 2015;16(5):11728–11749.

Venkatadri R, Muni T, Iyer AK, Yakisich JS, Azad N. Role of apoptosis-related miRNAs in resveratrol-induced breast cancer cell death. *Cell Death Dis.* 2016;7:e2104.

Wang L, Wei D, Han X, Zhang W, Fan C, Zhang J, et al. The combinational effect of vincristine and berberine on growth inhibition and apoptosis induction in hepatoma cells. *J Cell Biochem.* 2014;115(4):721–730.

Wang X, Wang N, Li H, Liu M, Cao F, Yu X, et al. Up-regulation of pai-1 and down-regulation of upa are involved in suppression of invasiveness and motility of hepatocellular carcinoma cells by a natural compound berberine. *Int J Mol Sci.* 2016;17(4):577.

Wang J, Qi Q, Feng Z, Zhang X, Huang B, Chen A, et al. Berberine induces autophagy in glioblastoma by targeting the AMPK/mTOR/ULK1-pathway. *Oncotarget.* 2016;7(41):66944–66958.

Wang F, Zhao J, Liu D, Zhao T, Lu Z, Zhu L, et al. Capsaicin reactivates hMOF in gastric cancer cells and induces cell growth inhibition. *Cancer Biol Ther.* 2016;17(11):1117–1125.

Wilkinson S, Farrelly S, Low J, Chakraborty A, Williams R, Wilkinson S. The use of complementary therapy by men with prostate cancer in the UK. *Eur J Cancer Care (Engl).* 2008;17:492–499.

World Health Organization (WHO). 2016a. Available from: http://www.who.int/cancer/en/. Accessed January 1, 2016.

World Health Organization (WHO). 2016b. Available from: http://www.who.int/medicines/areas/traditional/definitions/en/. Accessed January 10, 2016.

Wu SH, Hang LW, Yang JS, Chen HY, Lin HY, Chiang JH, et al. Curcumin induces apoptosis in human non-small cell lung cancer NCI-H460 cells through ER stress and caspase cascade- and mitochondria-dependent pathways. *Anticancer Res.* 2010;30(6):2125–2133.

Wu J, Xu J, Han S, Qin L. Effects of genistein on apoptosis in HCT-116 human colon cancer cells and its mechanism. *Wei Sheng Yan Jiu.* 2014;43(1):1–5.

Wu L, Guo L, Liang Y, Liu X, Jiang L, Wang L. Curcumin suppresses stem-like traits of lung cancer cells via inhibiting the JAK2/STAT3 signaling pathway. *Oncol Rep.* 2015;34(6):3311–3317.

Xiao X, Liu Z, Wang R, Wang J, Zhang S, Cai X, et al. Genistein suppresses FLT4 and inhibits human colorectal cancer metastasis. *Oncotarget.* 2015;6(5):3225–3239.

Yang CS, Maliakal P, Meng X. Inhibition of carcinogenesis by tea. *Annu Rev Pharmacol Toxicol.* 2002;42:25–54.

Yang CL, Ma YG, Xue YX, Liu YY, Xie H, Qiu GR. Curcumin induces small cell lung cancer NCI-H446 cell apoptosis via the reactive oxygen species-mediated mitochondrial pathway and not the cell death receptor pathway. *DNA Cell Biol.* 2012;31(2):139–150.

Yang YM, Yang Y, Dai WW, Li XM, Ma JQ, Tang LP. Genistein-induced apoptosis is mediated by endoplasmic reticulum stress in cervical cancer cells. *Eur Rev Med Pharmacol Sci.* 2016;20(15):3292–3296.

Yu R, Zhang ZQ, Wang B, Jiang HX, Cheng L, Shen LM. Berberine-induced apoptotic and autophagic death of HepG2 cells requires AMPK activation. *Cancer Cell Int.* 2014;14:49.

Zhang Y, Ge Y, Chen Y, Li Q, Chen J, Dong Y, Shi W. Cellular and molecular mechanisms of silibinin induces cell-cycle arrest and apoptosis on HeLa cells. *Cell Biochem Funct.* 2012;0(3):243–248.

Zi X, Agarwal R. Silibinin decreases prostate-specific antigen with cell growth inhibition via G1 arrest, leading to differentiation of prostate carcinoma cells: Implications for prostate cancer intervention. *Proc Natl Acad Sci U.S.A.* 1999;96(13):7490–7495.

Zi X, Grasso AW, Kung HJ, Agarwal R. A flavonoid antioxidant, silymarin, inhibits activation of erbB1 signaling and induces cyclin-dependent kinase inhibitors, G1 arrest, and anticarcinogenic effects in human prostate carcinoma DU145 cells. *Cancer Res.* 1998;58(9):1920–1929.

Index

Milton Keynes UK
Ingram Content Group UK Ltd.
UKHW040054071024
449327UK00019B/555

9 780367 657338